Colonic Drug Absorption and Metabolism

DRUGS AND THE PHARMACEUTICAL SCIENCES

A Series of Textbooks and Monographs

edited by

James Swarbrick
School of Pharmacy
University of North Carolina
Chapel Hill, North Carolina

Colonic Drug Absorption and Metabolism

edited by
Peter R. Bieck

Human Pharmacology Institute
Ciba-Geigy GmbH and
University of Tübingen
Tübingen, Germany

Marcel Dekker, Inc.   New York • Basel • Hong Kong

Library of Congress Cataloging-in-Publication-Data

Colonic drug absorption and metabolism / edited by Peter R. Bieck.
     p.   cm. -- (Drugs and the pharmaceutical sciences ; 60)
    Includes bibliographical references and index.
    ISBN 0-8247-9013-8 (alk. paper)
    1. Colon (Anatomy)--Effect of drugs on. 2. Drugs--Metabolism.
I. Bieck, Peter R. II. Series: Drugs and the
pharmaceutical sciences ; v. 60.
    [DNLM: 1. Colon--drug effects. 2. Colon--metabolism. 3. Drugs-
-metabolism. W1 DR893B v.60 1993]
PM355.C65 1993
615'.7--dc20
DNLM/DLC
for Library of Congress   93-17265
                                              CIP

The publisher offers discounts on this book when ordered in bulk quantities. For more information, write to Special Sales/Professional Marketing at the address below.

This book is printed on acid-free paper.

MARCEL DEKKER, INC.
270 Madison Avenue, New York, New York 10016

Current printing (last digit):
10 9 8 7 6 5 4 3 2 1

PRINTED IN THE UNITED STATES OF AMERICA

Preface

In recent years, many pharmaceutical prolonged-release preparations have been developed that deliver drugs along the entire intestine and to the colon. Yet, little is known regarding the extent to which drugs are absorbed by the colonic mucosa. This book was written to provide the basis for optimal design of drug delivery to the colon.

Our goal is to present an integrated series of chapters on the anatomical and physiological basis of drug absorption and metabolism in the colon. The contributors, all knowledgeable researchers and clinicians throughout the world, provide an excellent summary of the current knowledge in this area. Chapter 1 discusses the various physiological functions and properties of the colon that may influence drug absorption. Chapter 2 deals with reactions mediated by the enzymes within the intestinal mucosa and the gut flora, as well as the pharmacological and toxicological importance of biotransformation in the colon. Chapter 3 explores in vitro models of drug transport and metabolism of colonic tissue, cellular, and subcellular preparations. Chapter 4 describes scintigraphic techniques and applications in colonic drug absorption, including monitoring the transit of food and investigating the behavior of pharmaceutical dosage forms within the gastrointestinal tract.

Chapter 5 illustrates the techniques for studying colonic drug absorption, including intubation, remote control devices, and colonoscopy. Chapter 6 offers a comparison of the in vitro results of drug dissolution with the in vivo profile of dissolution or absorption and evaluates the efficacy of controlled-release formulations. Chapter 7 reviews dosage forms for drug delivery to the colon, with emphasis on osmotic delivery systems and their value. Chapter 8 details the scientific challenge of developing convenient and effective delivery systems for peptide and protein drugs. Chapter 9 assesses the impact of disease on colonic drug absorption and explores the effects of physiochemical and physiological variables on mucosal permeation. The last chapter summarizes current knowledge about drug-induced injuries to the colon, focusing on both common and rare causes, and luminal adverse effects on the colonic mucosa.

This book will be of value to biomedical researchers and specialists working in the field and will improve their understanding of the absorptive and metabolic function of the human colon.

I am most grateful to the many distinguished scientists and clinicians from different disciplines with a common interest in colonic drug absorption who participated in the creation of this text. I believe that it provides a comprehensive summary, in a manner understandable to pharmacists, pharmaceutical and biomedical researchers, gastroenterologists, and graduate students in these disciplines.

I dedicate this book to my former teachers and colleagues Professor W. Dolle and Dr. H. Hess, and to my wife Inge. She helped to prepare the index and has made personal sacrifice and shown great tolerance.

Peter R. Bieck

Contents

Contributors

Karl-Heinz Antonin, M.D. *Human Pharmacology Institute, Ciba-Geigy, Tübingen, Germany*

Dierk Brockmeier, Ph.D. *Department of Biometrics and Documentation, Clinical Research, Hoechst AG, Frankfurt, Germany*

Terry L. Burkoth, Ph.D. *Alza Corporation, Palo Alto, California*

Pierre Dechelotte, M.D.* *Medizinische Hochschule, Hannover, Germany*

Christine A. Edwards, Ph.D. *Department of Human Nutrition, Glasgow University, Glasgow, Scotland*

---

* Present affiliation: Policlinique Hôpital Charles Nicolle, Rouen, France

Johann W. Faigle, Ph.D.* *Research and Development Department, Pharmaceuticals Division, Ciba-Geigy Limited, Basel, Switzerland*

Deborah A. Fox, Ph.D. *Alza Corporation, Palo Alto, California*

Christoph H. Gleiter, M.D. *Human Pharmacology Institute, Ciba-Geigy, Tübingen, Germany*

John G. Hardy, Ph.D. *Pharmaceutical Profiles Limited, Nottingham, England*

Martin Mackay, Ph.D. *Drug Preformulation and Delivery, Ciba-Geigy Pharmaceuticals, Horsham, West Sussex, England*

Stuart A. Riley, M.D. *Department of Internal Medicine, Northern General Hospital, Sheffield, England*

Michael F. Schwenk, M.D., Ph.D.† *Medizinische Hochschule, Hannover, Germany*

Felix Theeuwes, D.Sc. *Alza Corporation, Palo Alto, California*

Eric Tomlinson, D.Sc., Ph.D.‡ *Somatix Therapy Corporation, Alameda, California*

Patrick L. Wong, Ph.D. *Alza Corporation, Palo Alto, California*

---

* Retired
† *Present affiliation*: Department of Environmental Toxicology, Federal Health Agency, Stuttgart, Germany
‡ *Present affiliation*: GeneMedicine, Inc., Houston, Texas

Colonic Drug Absorption and Metabolism

1

Anatomical and Physiological Basis: Physiological Factors Influencing Drug Absorption

Christine A. Edwards
Glasgow University, Glasgow, Scotland

I. INTRODUCTION

Interest in the physiology and function of the human colon has increased over the last two decades mainly in relation to the bacterial metabolism, and subsequent absorption or action, of carbohydrate and dietary fiber, and a vast array of xenobiotic and intrinsic molecules. Although the colon does not compare well with the small intestine, it does have a significant absorption capacity and since the residence of material in the human colon is 2–3 days rather than the 5 hr in the small intestine, there is a large potential for the slow absorption of drugs and other materials. This is especially relevant to the recent use of prolonged-release pharmaceutical dosage forms that deliver drugs slowly throughout the gut and may reach the colon before delivering much of their dose. In this chapter, the physiological functions and properties of the colon that may influence the absorption of drugs and other molecules will be discussed.

II. ANATOMICAL AND PHYSIOLOGICAL REGIONS

In terms of size and complexity, the human colon falls between that of carnivores such as the ferret, which has no identifiable junction between ileum and colon, and herbivores such as the rat, which has a voluminous cecum (1). The human cecum is small and there is a rudimentary appendix. The human colon can be divided into three functional areas; the cecum and proximal colon, which act as a fermentation chamber; the transverse colon, the motor patterns of which may hold material in the proximal colon or propel it distally but that may also be an important site for the absorption of water; and the rectum, which acts as a reservoir for fecal material and allows defecation to be delayed until socially convenient.

The physiology of the proximal and distal colon differs in several respects that relate to their function (Table 1) and may affect drug absorption at each site. The physical properties of the luminal contents of the colon also change, from liquid contents in the cecum to semisolid contents in the distal colon. In addition to the site of the colon, there may also be differences in the environment of a

Table 1 Major Differences Between Proximal and Distal Colon

	Proximal	Distal
Function	Fermentation chamber, absorption	Absorption, storage
Innervation	Vagal/pelvic Splanchnic/lumbar Muscle more distensible	Pelvic, lumbar, greater sensitivity to neural stimulation
Blood supply	Superior mesenteric artery and vein, greater blood flow	Inferior mesenteric artery and vein
Absorption	92% of chloride-dependent transport is electroneutral; greater overall capacity	Chloride-dependent transport mainly, amiloride sensitive
Luminal contents	Liquid, lower pH (4.6–7.8), higher SCFA, very active bacterial metabolism	Semisolid, neutral pH, lower SCFA, lower bacterial activity

drug or molecule depending on whether it is in the bulk phase or next to the mucosa, and whether it is free in the aqueous phase or bound to, or trapped in, solid material such as dietary fiber residues.

III. INNERVATION

The colon is supplied with both sympathetic and parasympathetic nerves but the sympathetic nerves dominate, maintaining a tonic inhibition of colonic motor activity (2). The parasympathetic supply is via the vagus and pelvic nerves. The vagus supplies the proximal and up to the midpoint of the transverse colon, whereas the pelvic or sacral spinal nerves appear to innervate the entire colon (3). Stimulation of the parasympathetic nerves causes an increase in contractility, but the proximal colon contracts rhythmically whereas the distal colon undergoes tonic contraction and shortening (2). The sympathetic supply is from the postganglionic neurons from the inferior mesenteric ganglion, which arise from the lumbar preganglionic outflow from the 2nd, 3rd, and 4th ventral lumbar roots. There is some input from the splanchnic system. The splanchnic influences the proximal colon whereas the lumbar system influences the whole colon. The α-adrenergic receptors are stimulatory and β-adrenergic receptors are inhibitory (4). Electrical stimulation of the sympathetic nerves decreases the effects of parasympathic stimulation and inhibits spontaneous motility. Blockade of α_2-adrenergic receptors or administration of phentolamine increases colonic motility (5).

In addition to the autonomic nervous system, as in all regions of the gut, there is a complex enteric nervous system consisting of two major plexuses, myenteric and submucosal, and other small plexuses in the mucosal and muscle layers. Each plexus contains organized networks of ganglia that process signals between cells in the same plexus, with other plexuses, extrinsic nerves, and the muscle and mucosal layers, playing a major role in the control of motility and transport processes. In the proximal colon the myenteric plexus is regular and stellate. The distal colon has a unique pattern of shunt fascicles (6) (large nerve bundles running along the axis of the colon) that connect with the ganglia and originate

from the pelvic nerve (7). The rectal plexus has an irregular appearance (6). The distal colon is more sensitive to neural stimulation,
having enhanced cholinergic activation and increased muscle sensitivity to acetylcholine (8) in comparison with the proximal colon.

Central neural control of the colon is demonstrated in the effects
of stress on colonic motility, which are discussed below.

The detailed organization and neurotransmitters involved in
this network of neural control are beyond the scope of this chapter
but are well reviewed elsewhere (9).

IV. MOTILITY

Studies of colonic motility in vivo usually rely on measurement of
changes in muscle electrical activity that may determine contractions, manometric measurements of changes in colonic pressure
caused by contractions, or strain gauges to measure contractions
more directly. All approaches provide useful information but when
used separately may not give a complete picture of colonic motor
events. Electrical activity may not produce measurable contractions (9) and manometric techniques can only detect contractions
that occlude the lumen sufficiently to register as an increase in
pressure (this is particularly important where the lumen is large
and pressure increases are dampened). In vitro measurements
using strips or segments of colon may suggest mechanisms and
patterns of electrical and motor activity, but their role must be
assessed in vivo in an intact colon with enteric and autonomic
nervous system (ANS) and central nervous system (CNS) connec-
tions maintained. Since in vivo studies in humans involve intubation and often bowel cleansing (sometimes with cathartics that
may sensitize the colon; 10) it is difficult to assess whether the
same patterns would be seen without the invasive tubes and with
a colon full of chemically and mechanically stimulating contents.

A. Muscles

The smooth muscle is composed of an inner continuous circular
and an outer longitudinal layer. The longitudinal muscles are com-

prised of three strands (taeniae coli) that fuse together in the rectum.

The mechanical properties of the proximal and distal colon differ. Human colonic muscle taken at surgery from the right colon was more distensible than that of the left colon and the maximum spontaneous active stress was exerted at greater degrees of stretch than those of the left colon (11). The sigmoid colon produced the most powerful contractions in vitro, but the contractile frequency of the circular muscle of the right colon was greater than that of the left colon (12).

B. Electrical Basis of Motility

Contraction of the colonic muscles is determined by the electrical activity of the muscle cells (myogenic control; Table 2). Periodic variations in the resting membrane potential (control activity) determine the excitability of the muscle and the timing of contractions. When the membrane depolarizes beyond a threshold potential, the muscle generates an electrical response activity (ERA) that results in muscle contraction. Whether an ERA occurs or not is dependent on a neural or chemical signal occurring at the time of depolarization (9). The degree of co-ordination and organization of the electrical activities and hence contractions is dependent on the coupling between adjacent muscle cells. This coupling may not be as efficient in the colon as in the small intestine, producing a disorganized or poorly phase-locked control activity and disorganized and short-duration contractions. This may also be related to a lack of gap junctions (12).

Two types of control activity occur: electrical control activity (ECA) and contractile electric complex (CEC). Electrical control activity, often called slow waves, is spontaneous, continuous (14, 15), and has a frequency in the range of 2–13 cycles/min (16). In the proximal colon these oscillations in membrane potential appear to migrate orad (17), but from the midcolon they spread aborad. They are very small in the rectum. The slow wave can occur at a variety of frequencies depending on the level of stimulation (16). Recorded slow-wave frequency in vivo may be the result of summation of different frequencies (15).

The ERA activity responsible for short-duration contractions that superimpose on the ECA when stimulated at depolarization is

Table 2   Electrical and Contractile Motility Patterns in the Colon

ECA (slow wave)	Slow variations in resting membrane potential of muscle.	⎫
CEC (oscillatory potential)	25/40 cycles/min; may originate in longitudinal muscle associated with long-duration contractions.	⎬ Control activity
SSB (discrete electrical response activity)	Burst of activity lasting <5 sec. These may relate to nonpropagating contractions that may act as baffles to hold up flow along colon and mix luminal contents.	⎫
LSB (continuous electrical response activity)	Long-lasting activity (>10 sec) may relate to propagating motor activity.	⎬ Response activity
Unknown	Determines "mass movements" or high-amplitude contractions.	⎭

known as short spike bursts SSB (discrete electrical response) (19). They last less than 5 sec and occur in series (19). These are thought to correspond to nonpropagating haustral contractions and are most predominant in the proximal colon (19).

The second type of control activity, the contractile electrical complex, is the oscillatory membrane potential that appears to be associated always with long-duration contractions (20). These oscillations are intermittent and may require neural or chemical stimulation. They have a higher frequency than the ECA (slow waves); 24–36 cpm (14). They probably originate in the longitudinal muscle (20, 21) but may need the circular muscle layer to be present (22). They are inhibited by atropine and isopropyl norepinephrine (19), unlike the ECA or slow waves.

If neural or chemical signals occur when the CEC is below threshold a different form of ERA occurs; long spike bursts (LSB or continuous electrical response) that last more than 10 sec and are associated with long-duration contractions that may propel contents along the colon (19).

The electrical activities of the longitudinal and circular muscle layers appear to be uncoupled but may be phase-locked during stimulation (23, 24). The electrical activity associated with mass movement or high-amplitude contractions (see below) is not yet known (9).

C. Motility Patterns

To fulfill the roles of fermentative chamber, propulsion, and reservoir, the different regions of the colon have different predominant motor patterns. Because of the inaccessibility of the human colon and the necessary invasiveness of the techniques used it has been difficult to assess the role of different motility patterns and their relation to flow. Most colonic contractions consist of ringlike contractions of both longitudinal and circular muscle, causing the relaxed areas to bulge outwards in haustra. These may move forwards or backwards or remain stationary. The proximal colon has mainly retropulsive and mixing movements. The sigmoid colon has a degree of tonic and phasic motor activity and may hold colonic contents back in the descending colon to allow water absorption.

Long-term manometric studies of colon motor activity in humans (25–27) have revealed the major stimuli of motor activity. In these studies manometric probes were placed in the colon at colonscopy either in the rectum (26), up to the transverse (25), or through to the ascending colon (27). The probes were left in place and recorded from for 24 hr. All studies noted that the colon was very quiescent during sleep, at night or during the day. In a further short-term study there was an increase in the motility index of the colon between rest and vigilance maintained by conversation (28). The major stimuli for colonic motility in the long-term studies were awakening and meals. Narducci and colleagues (25) recorded mainly low-amplitude contractions that occurred singly or in bursts but showed no recognizable pattern. High-amplitude contractions were also seen propagating mainly over long distances at approximately 1 cm/sec and were often associated with an urge to defecate. These "mass movements" or high-amplitude propagated contractions were also reported by Kumar et al. (26) in the rectum and throughout the colon by Bassotti and Gaburri (27). They are thought to occur by relaxation of the gut caused by descending neural inhibition and propagation of one ringlike contraction through colonic shortening starting in the ascending or midtransverse colon. Propulsion can also occur with the long-duration contractions associated with LSB, although aboral movement can occur during short- and long-duration contractions. The nonpropagating contractions associated with SSB appear to act as baffles slowing transit through the intestine. In diarrhea SSB activity is virtually absent (29) but the motility is very active in constipation (20). This presents an apparent paradox in the colon in that unless propagation is measured, an increase in colonic activity may in fact reflect a slowing of transit, and where there is an inhibition of motility this may be related to an increased flow and decreased transit time.

Other studies of colonic motility in the human have mainly concentrated on the sigmoid colon and rectum because of the problems of accessibility.

D. Stimulus for Colonic Motility

Various extrinsic stimulants of colonic motility have been reported (Table 3). The most powerful after awakening (25–27) is the gastrocolic or meal response (30).

The motility index of the colon (area under motility trace or product of mean amplitude of the pressure waves multiplied by the sum of the duration of each pressure wave) is increased after ingestion of a high-calorie meal (1000 kcal; 30). Very little stimulation is seen after a 350 kcal meal. "Mass movement" contractions are not stimulated by a meal.

There have been many studies of this phenomenon and a high-calorie meal is often used as the stimulus for the study of other factors influencing colonic motility. The mechanism for this response is not clear although it may involve hormonal (30–36) and neural elements (31, 34). Vagal efferent pathways appear to be involved (35), but the role of pelvic nerves is unclear (35, 37). Colonic luminal pressure may modulate the gastrocolic response. Semba showed that when intracolonic pressure is low the colon does not

Table 3 Possible Stimulants of Motor Activity

		References
Extracolonic		
Awakening		25–27
Gastrocolonic reflex	High-calorie meal (1000 kcal). Two phases: early <30 min after meal may relate to food in stomach and duodenum. Late: 60–90 min after meal. May relate to food entering colon	30–37
Specific food components	Coffee	53
Stress	May provoke or inhibit motility depending on type of stress: mental, physical, or emotional	54–57
Intracolonic		
Distention	Caused by excessive fluid flow, gas production	59
Endogenous molecules	Bile acids	60, 61
Bacterial products	Short chain fatty acids? Hydroxy fatty acids, gas	59, 63–65
Mechanical stimulation	Particles of fibers such as wheat bran	66–69

respond to gastric distention but does if colonic pressure is high (37). Physical exercise may also enhance the colonic response to meals (38). It has been suggested that like other intestinal responses to food that there is a cephalic, gastric, and intestinal phase (32), with colonic motility increasing at the sight and smell of food, the presence of food in the stomach, and presence of food in the intestine, both small and large. The cephalic phase is difficult to demonstrate (39–41), perhaps because the effect is short-lived in humans and we quickly adapt to the presence of food. The colonic response to food is made up of two components. The first is a rise in motility seen very quickly within 30 min of the meal, which must be related to food in the stomach or duodenum (30). Many studies have re-

ported an immediate increase in colonic motility when food enters
the stomach (30, 41–44) but there is an independent action of food
in the small intestine. Colonic motor activity is stimulated by fat
infusion directly into the duodenum, but not intravenous fat (45);
and colonic stimulation is also seen when magnesium sulfate,
amino acids, and sodium oleate are introduced in the duodenum
(46, 47).

There is a second increase in colonic motility about 70–90 min
after the meal (30). In the dog there appear to be three phases, with
an increase in motility at 5 min, before 2 hr, and 8 hr after the meal
(48). This late response may be related to the entry of the meal
into the colon (48). The stimulus in each region of the gut may be
mechanical or chemical, but in light of the inconsistent effect of
gastric distention (49, 50) is more likely to be via chemoreceptors
(32). Most of these studies have concentrated on the distal colon.
Studies that have measured motility in more proximal colon have
shown differences in the meal response. In the dog the early postprandial phase was seen only in the distal colon and not the proximal, but the later phases were seen throughout the colon (48). Responses in the mid- and transverse colon of the human colon have
been observed (25, 51) and a small response seen in the proximal
colon (51) in one study, and to solid meals and not liquid meals in
another (52).

In relation to food intake, a recent study (53) showed an increase
in distal colonic motility in preselected "responders" after ingestion of coffee, both normal and decaffeinated, but not after ingestion of hot water.

The other major general stimulant of colonic motility is stress.
Interest in the effects of stress on the colon in humans has centered
around studies of the irritable bowel syndrome (IBS) (54). The effect of stress is very difficult to study and many different types of
stress have been applied experimentally to humans. None of these
may relate to the stressful situations in real life that result from
emotional or mental pressures (54). Many studies involve physical
stress, such as immersing a hand in water at 4°C or instilling water
at 4°C into the external auditory meatus of one ear. Mental stress
tests involve tasks performed during distracting aural or visual
stimuli, such as two different audio tapes played simultaneously
in either ear (dichotomous listening test). Several groups have
shown an increase in colonic motility in response to physical stress

(28, 55–57). Narducci et al. (55) found an increase in colonic motility in subjects given mental stress in the form of "ball sorting" and Stroop test tasks (subjects read in fast sequence 50 cards on which the name of a color is written in another color). Erkenbrecht et al. (56), however, found no effect of mental stress on unstimulated colonic motility and a decrease in motility patterns after a meal when mental stress was applied as a dichotomous listening test.

The most relevant studies are still those carried out by Almy et al. (57), who monitored colonic motor activity and appearance during interviews with patients with IBS about life events that the patients found very emotionally stressful. They recorded increased colonic motility during the time talking about the stressful life event compared to discussion of neutral topics.

The intraluminal factors that influence colonic motility are particularly poorly understood. Distention by the luminal contents or gas produced by bacterial fermentation may induce propulsive activity (58, 59). Bile acids and hydroxy fatty acids have also been shown to stimulate propulsion (60–62). The fermentative function of the proximal colon has encouraged speculation that the short-chain fatty acids (SCFA) produced by the bacteria may influence motility, perhaps to prevent movement of material from the fermentation chamber and allowing bacteria to maintain adequate populations. However, SCFA have been shown both to stimulate (63) and inhibit (64) colonic motility in the rat in vitro and had no effect on proximal colonic capacitance in the human (65). Particulate matter such as dietary fiber residues like bran (66), or indeed plastic pellets (67, 69), also stimulate motility and produce more frequent and more liquid stool (66–69), perhaps by stimulation of mechanoreceptors. The relationship of other physical properties of luminal contents, such as viscosity and density, and colonic motility have not been studied in any detail but need investigation.

A. Transit Time and Residence

One of the major determinants of the absorption of a compound from the colon is the residence in any particular segment of the colon. The time taken for food to pass through the colon accounts for most of the time the food is in the gut. In normal subjects this is about 78 hr for expulsion of 50% ingested markers (70) but may range from 18 to 144 hr. Steady-state measurement of mean transit time after ingestion of markers for several days gave a mean transit time of 54.2 hr (71). Measurement of transit through each section of the colon may be measured by ingestion of markers and subsequent x-ray (72) or gamma scintigraphy after oral ingestion (73) or perfusion into the cecum (65, 74). Segmental transit measured by radioopaque markers gives values of 35 hr for the whole colon and 11.3 hr, 11.4 hr, and 12.4 hr for the right, left, and rectosigmoid colon (72). Men had slightly shorter transit times than women, and this was most apparent in more proximal colon (72). X-ray studies have the advantage that the colon is full of solid material and the subjects are on a normal diet. Scintigraphic studies in which the label is ingested in a meal also involve patients with a full colon but require higher doses of radiation if a short half-life isotope is used and complex analysis to compensate for overlap of small and large intestines (73). Studies in which the radiolabel is infused directly into the colon allow more detailed measurements of colonic flow and transit but are usually performed with a prepared or washed out colon, so that the transit time of the contents is mainly that of the liquid infusate and may not relate to more solid material.

Krevsky et al. (74) injected a bolus of 8 ml and monitored the colon for 48 hr. Most of the label had been defecated by this time. They reported a rapid emptying of the proximal colon (88 min) and suggested that the transverse colon was a major storage site. Gamma scintigraphy can be carried out simultaneously with intraluminal pressure studies so that flow can be related to motility. In one study, eating a high-calorie meal increased nonpropagating contractions mainly in the descending colon and the radiotracer moved from the splenic flexure to the transverse and sigmoid colon (75). Another approach is to measure colonic capacitance (65) by infusing radiotracer and measuring the volume of different areas of the colon. Using this technique Kamath et al. (65) showed that the right colon acted as a reservoir but the capacitance was reduced by oleic acid.

1. Factors Affecting Transit Time

Transit through the colon is mainly affected by the diet. A low-residue diet is associated with slow transit and small fecal output (76). A high mixed-fiber diet or a diet supplemented with wheat bran or ispaghula decreases the transit time and increases stool output (76–78). This may be associated with an increase in LSB and a decrease in SSB (63). The mechanism by which dietary fiber speeds transit through the colon is not well understood. There are probably several mechanisms and each fiber has its own idiosyncratic action. Fibers with a large water-holding capacity, that is retained after fermentation by the rich colonic flora, may just increase the bulk of colonic contents (79) and act by distention (59). Gas produced by fermentation may also act this way. The physical presence of dietary residue may stimulate mechanoreceptors, as described above (66, 67, 69). The best stool bulkers are the least fermented. However, fermentation studies of different dietary fibers in vitro compared with transit measurement in humans who had ingested the same fibers and who provided the inoculum for the in vitro studies showed that transit time can be accelerated without an increase in stool output and that fermentation was necessary for acceleration of transit time (80). The mechanism for this is not known. The factors determining segmental transit are particularly poorly understood and more work is needed here.

Transit time is slower in women than men and is related to the circulating steroid hormones in the rat, but the menstrual cycle does not seem to have an effect in women (81, 82). Drugs and disease may affect transit. This is particularly important in colonic disease, where colonic administration of drugs may be necessary. This will be discussed in another chapter.

2. Transit Time of Pharmaceutical Dosage Forms Through the Intestine

Drug delivery to the colon, if from an oral route, depends on gastric emptying and small bowel transit time. The transit of various dosage forms through the small intestine was monitored after ingestion with and without food (83). Solutions and pellets emptied from the stomach with the food but single units were retained in the stomach for some time. However, intestinal transit times were independent of dosage or food. The factors affecting the movement of pharmaceutical dosage forms through the upper bowel have been extensively reviewed elsewhere (84). When the dosage forms reach the colon, the transit depends on size. Small particles pass through the colon more slowly than large units (85, 86) but the density and size of larger single units had no real effect on colonic transit. In a recent study (87) 50% of large units reached the splenic flexure within 7 hr of entering the colon independent of density. Larger units had a tendency to travel more quickly, but this was not biologically significant. Similar results were gained in a study of patients with IBS (88).

Drugs administered to the colon by enema may have a limited spread, reaching mainly the distal and midtransverse colon (89). Spreading of the enema solution was directly related to the level of colonic motility (89), although there was no significant effect of eating a meal.

V. ABSORPTION AND SECRETION

A. Mucosal Structure

The absorptive capacity of the colon is much less than that of the small intestine, and this is mainly due to a lower surface area. The mucosal surface of the colon at birth is similar to that of the small intestine, but rapidly changes with the loss of the villi leaving a flat mucosa with deep crypts (90). As the gut ages there is a decrease in the number of nongoblet crypt cells and this is related to an increase in fecal water (91). Despite the lower surface area, the colon has a large capacity for the absorption of water, electrolytes, and the short-chain fatty acids formed during fermentation of carbohydrate by the colonic bacteria. Of about 1500 ml water entering the colon, only 200 ml a day is excreted. The colon is capable of absorbing 4 L of water a day and can withstand an infusion rate of 6 ml/min before there is an increase in fecal water (92, 93). The colon can also absorb 816 mEq sodium and 44 mEq potassium per day (92). There are some differences in the absorptive capacity and the control of absorption mechanisms in different parts of the colon and so residence time in a particular zone may be important.

B. Transport Studies and Mechanisms

1. Electrolytes

The human colon absorbs sodium and chloride and secretes bicarbonate and potassium against electrochemical gradients (Table 4). Sodium absorption and potassium secretion are increased by aldo-

Table 4   Colonic Absorption of Electrolytes and Other Molecules

Sodium Proximal colon	50% dependent on Cl^- of which 8% electrogenic and amiloride sensitive and 92% electroneutral (parallel Na^+/H^+; Ca^{2+}/HCO^{3-} exchange). Other active mechanism non-chloride-dependent unknown.
Distal colon	50–100% amiloride sensitive, increased by aldosterone
Potassium	Active secretion dependent on Na^+/K^+-ATPase on basolateral membrane. Absorption by passive diffusion by paracellular and transcellular routes. Secretion stimulated by aldosterone.
Bicarbonate	Secretion in exchange for Cl^-, formed by carbonic anhydrase on epithelial surface.
Chloride	Coupled to Na^+ and bicarbonate.
Water	Follows Na^+ and SCFAs by solvent drag.
SCFA	Passive absorption, probably in un-ionized form, may depend on site in colon.
Organic acid and bases	Dependent on lipid solubility and relation to pKa to pH microclimate.
Sugars Amino acids	Poorly absorbed by passive diffusion

sterone. All regions of the colon actively absorb sodium and chloride at more or less similar rates in vitro (94), although some researchers report a higher absorption in the transverse colon (95) and in vivo perfusion and dialysis studies indicate a greater absorption in the proximal and transverse colon (96–99). In the human 50% of active sodium transport is chloride dependent (94). This is different from the rat, in which active sodium transport is 80% dependent on chloride (99, 100). The mechanism of absorption differs with site. As one moves distally in the colon, electrogenic sodium transport (blocked by amiloride) increases (94). In the proximal colon it is a minor pathway (about 8%; 101) but in the distal colon it accounts for 50–100% of the sodium transport (94, 102). The other major pathway is an electrically neutral chloride-coupled cotransport, most likely a parallel sodium–hydrogen, chloride–bicarbonate exchange (100, 101). The third active transport that is not chloride dependent and not blocked by amiloride (94) is responsible for 7–15% of the transport in the distal colon and approximately 50% in the proximal colon. The exact pathway is not known. Bicarbonate secretion involves the formation of bicarbonate by carbonic anhydrase on the epithelial surface (103). Bicarbonate in the lumen neutralizes the acidity caused by short-chain fatty acid production.

Potassium transport is also different in the proximal and distal colon. The proximal colon seems to secrete potassium in vitro, whereas the distal colon absorbs potassium under basal conditions (104). Aldosterone-treated tissue however, secretes potassium from both sites (104). In the presence of SCFA, the rabbit proximal colon absorbed potassium (105). Potassium secretion in both the proximal and distal colon is an active process driven by the oubain-sensitive Na^+/K^+-ATPase on the basolateral membrane. In vitro studies of rabbit colon (106) showed that when chloride was removed from the bathing solution in the distal colon, serosal to mucosal K^+ flux was reduced, increasing mucosal to serosal flux and resulting in K^+ absorption. Removing the chloride had no effect in the proximal colon. The mechanism is unknown but again emphasizes the segmental heterogeneity of the proximal and distal colon. Potassium absorption is by passive diffusion, probably by a paracellular route (107), although some observations suggest a transcellular route (108). The absorption and secretion of K^+ are further complicated by a low potassium concentration in the juxtamucosal microclimate which is more stable in the proximal colon than the distal colon (107, 109). This difference may relate to a higher transcellular route of potassium transport in the distal colon. The absorption of electrolytes is influenced by activity of the mucosal plexus (110, 111).

2. Water Absorption

The colon is a major site of water absorption and forms a semisolid stool from the liquid ileal effluent. The water is absorbed mainly in the proximal colon, but the site of absorption may be delayed by the presence of a slowly fermentable fiber of high water-holding capacity such as ispaghula (112). The absorption of water may be critical for the absorption of drugs. Water retained in the lumen may dilute the drug concentration but may also facilitate movement and mixing within the lumen to allow better contact with the epithelial surface. Water absorption occurs secondary to the absorption of electrolytes and SCFA by solvent drag. The colonic epithelium is less permeable than the small intestine and allows the water to be extracted against a large transepithelial osmotic gradients. There is evidence for hypertonic absorption across the crypts of the rat distal colon, where an osmotic pressure equivalent to 5 atm would be required to dehydrate the feces (113). Perfusion studies in the rat rectum of the promoting effect of sodium tauro-cholate and sodium ethylene diamine acetate (EDTA-Na) on anti-pyrine absorption (114) suggest that increasing water absorption with these two agents increased antipyrine absorption. This effect was significantly reduced by oubain, suggesting increased solvent drag by Na^+ absorption as the promotional mechanism.

3. Short-Chain Fatty Acids

Carbohydrate that enters the colon, in any form but mostly as dietary fiber or mucopolysaccharides, is fermented to short-chain fatty acids (SCFA) mainly acetic, propionic, and butyric. It is now well established that these SCFA are rapidly absorbed from the colon and are either used by the colonic epithelial cells (115, 116) or are transported to the liver. The mechanism of absorption in humans is thought to be by passive diffusion of the un-ionized form with luminal accumulation of bicarbonate (117, 118). However, by passing through a stable pH microclimate of about 6.8 at the mucosal surface, most of the SCFA will be in an ionized form (119) and this corresponds to the lack of effect of chain length and lipid solubility on the absorption of SCFA in the human rectum (119). This pH microclimate is achieved by the secretion of H^+ (120) produced by carbonic anhydrase on the mucosal membrane and may also be important in the absorption of drugs (see below). This corresponds to studies of the pony colon, where both SCFA absorption mechanisms occur in the proximal colon but SCFA are absorbed predominantly in the ionized form in the distal colon (121). The absorption of SCFA enhances the absorption of electrolytes and water (122), although some researchers dispute the effect on electrolytes in the rabbit cecum (123). Butyrate has been shown largely to reverse the effect of cholera toxin on water secretion in the rat in colonic loops in vivo (124).

4. Other Molecules and Drugs

Sugars such as glucose (125) and sucrose (126) are poorly absorbed in the adult human colon. Amino acids are passively absorbed by the horse colon (127). An in vitro study of the permeability of the colonic mucosa to oxalate and neutral sugars (128) suggests that the colon excludes molecules on the basis of size and charge. Bile acids and fatty acids increased absorption by increasing the transport sites but not the selectivity. Billich and Levitan (129) studied the effects of osmolality on net water movement in the human colon and, using mannitol and urea, estimated an equivalent pore size of 2.3 Å in comparison to the jejunum (8Å) and ileum (4Å) (130).

Lipid-soluble molecules are most readily absorbed by passive diffusion (131). Organic acids and bases and drugs are in general most rapidly absorbed in their lipid-soluble undissociated form.

Observations of the absorption of organic acids and drugs from the colon led Schanker (131) to suggest the existence of a more acid microclimate close to the mucosa. This has been confirmed by other workers in the guinea pig (132), rat (119, 132), and the human rectum (119). The pH measured at the mucosal surface of the human rectum was very stable and changed little despite large changes in luminal pH. The surface pH of 6.8 fell to only 6.26 when the luminal pH fell from 7.51 to 5.96. Thus the dissociation of a drug on absorption will depend on the relation of its pKa to the pH of the microclimate and not to the pH of the bulk phase. The microclimate thickness was about 840 μm and was dependent on the integrity of the mucous layer (132). The concentration of K^+ in the microclimate is also lower and independent of the luminal contents (109), being closer to serosal values. This low K^+ concentration is probably maintained by a high pre-epithelial diffusion barrier and a high paracellular shunt. It is more stable in the proximal colon than in the distal colon (107).

VI. MUCOSAL AND LUMINAL FACTORS AFFECTING ABSORPTION

As well as providing a stable pH environment, the mucous layer adjacent to the colonic mucosa also acts as a diffusion barrier (133). Smith et al. measured the movement of butyrate through the colonic mucus and compared it with movement through synthetic gels and the unstirred layer. They found no difference in the movement through mucus at different sites in the colon, but butyrate movement was only 50% of that through the unstirred layer and equivalent to that through an 8% polyacrilamide gel (133).

Mucus production in the colon is a function of goblet cells and since the proportion of goblet cells increases with age (although mainly due to a loss of other types of cell) this may be a factor that changes in the elderly (90). Mucins are degraded by the colonic bacteria (134). Thus changes in the intestinal flora may also affect the mucosal environment. The mucous layer may also be affected by disease and is thinned by the action of prostaglandins (119).

In the bulk phase, the physical properties of the luminal contents may be important in determining rates of absorption. The fluid contents of the cecum and proximal colon are progressively dehydrated until they are only 70% water in the rectum. This reduction in water content means there is less mixing in the bulk phase and thus less access to the mucosal surface. Factors that change the amount of water retained in the colon may increase the mobility and hence the absorption of molecules. However, many of the dietary fiber components with a higher water-holding capacity are fermented by the bacteria in the cecum and proximal colon and hence lose their water-holding properties (79). Those fibers with high water-holding capacity that are poorly fermented are often viscous (e.g., ispaghula) and this in itself will reduce mixing, although not to the extent of dehydrated feces. In the small intestine it is well established that viscous polysaccharides decrease the absorption of nutrients by inhibiting intraluminal mixing (135). This has not been investigated in the colon. Some dietary fibers such as pectin, have cation exchange properties that may bind charged molecules such as bile acids (136). This binding is increased at the low pH encountered in the colon (137) and may be a factor in the immobilization of some drugs. In addition, drug molecules could be trapped within the solid matrix of the concentrated dietary residue.

VII. COLONIC BACTERIA

The effect of the colonic bacteria on drug absorption by their metabolism of the drug molecules is discussed in another chapter. The bacterial fermentation of carbohydrate (dietary fiber or resistant starch) may, however, also influence drug metabolism. First by increasing the numbers of bacteria in the colon with the metabolic capability to transform and inactivate drugs (138, 139). Second the production of SCFA may have several indirect effects. The concomitant fall in pH (140, 141) may alter the dissociation of a drug and affect its binding to dietary residue. Ingestion of unabsorbed sugars such as lactulose (142) and dietary fiber (140, 141) has produced decreases in colonic pH; the greatest decreases are seen in the proximal colon (140, 143). The promotion of the absorption of water may increase drug absorption indirectly (114). The SCFA may influence colonic motility (63, 64) and transit time and also the water content of the stool (144). The SCFA may be trophic to the colonic epithelium, increasing the absorptive surface area (145). Finally, the SCFA may increase colonic blood flow (146).

VIII. BLOOD SUPPLY AND CIRCULATION

The absorption rate of drugs from the colon will be influenced by the rate of blood flow to and from the absorptive epithelium. Many factors will influence this blood flow, including the concentration of absorbed nutrients and changes in motor activity. The arterial blood supply to the proximal colon is from the superior mesenteric artery and the inferior mesenteric artery supplies the distal colon. The venous drainage is via the superior (proximal colon) and inferior (distal colon) veins. The arterioles and capillary branches pass to the epithelial surfaces between the crypts and form an extensive network of capillary plexi. The mucosal capillaries are fenestrated, the fenestral openings being most abundant in the upper third of the mucosa and on the side of the capillary wall facing the transporting epithelia. Measurement of blood flow to the colon is difficult and reported preprandial blood flow values range from 8 to 75 ml/min (147) due mainly to experimental differences. The proximal colon receives a greater share of the blood flow than more distal parts (148) and the total blood flow is less than that of the small intestine (147). Factors that increase blood flow include mechanical mucosal stimulation (149) and SCFAs (146, 150). Sympathetic stimulation decreases blood flow and parasympathetic stimulation increases it (147). Feeding appears to decrease blood flow 30–45 min after the meal (147). When the colonic muscle contracts there is an overall increase in blood flow to the colon to meet the demand of the contracting muscle, but mucosal capillaries may be constricted by the contraction and local blood flow reduced. The complex interaction between blood flow and motility has been extensively reviewed elsewhere (151).

IX. SUMMARY OF FACTORS AFFECTING ABSORPTION OF DRUGS FROM THE COLON

Before a drug is absorbed by the colonic epithelium it must reach the absorptive site (Table 5). Thus if given orally it must first reach the colon. Factors that affect transit time through the upper gastrointestinal tract and colonic motor patterns that influence residence and contact time will determine the site at which the drug is absorbed. The residence times in different parts of the colon may influence drug absorption because of the many physiological differences in the motility, luminal environment, absorptive capacity and other characteristics of the proximal and distal colon. Diet and the relation of drug ingestion to meal time may be very important. The drug must then move from the bulk phase to the epithelial surface. Factors such as colonic motor patterns, viscosity, binding, or entrapment will determine the rate of mixing and hence movement to the mucosal surface. The molecule must then pass across the mucous and unstirred layer to the membrane where it can pass

Table 5 Factors Affecting Drug Absorption in the Colon

Delivery rate to colon
Lipid solubility
Colonic residence: site of colon (proximal or distal) important
pH microclimate—pKa of drug
Mixing rate and resistance of contents: solidity or viscosity of content
Mucous barrier, unstirred layer
Mucosal pore size: permeability
Concentration: dilution by luminal contents, especially with high water-
  holding capacity dietary fiber
Bacterial metabolism

---

through either the lipid cell membrane or the water-filled pores. The thickness of the mucous layer may determine motility. The lipid solubility and size of the drug will determine ease of transport across the epithelium. The rate of absorption of weak bases and acids will be determined by the pH of the absorptive site, the pKa of the drug, and the oil/water coefficient. The pH microclimate at the mucosal surface may determine the dissociation of the drug and hence its absorption rate. The factors determining mucosal blood flow are not well understood but may be important. Finally, the metabolism of the drug by the intestinal bacteria and the changes in the luminal environment caused by fermentation have a major influence. The diet and especially the ingestion of dietary fiber and other bacterial substrates that are fermentable or that hold water in the colon may influence several of the steps before a drug is absorbed.

REFERENCES

1. C. E. Stevens, in *Physiology of Domestic Animals* (M. J. Swenson, ed.) Cornell University Press, Ithaca (1977).
2. H. Rostad, *Acta Physiol. Scand.* 89: 91 (1973).
3. L. Hulten, *Acta Physiol. Scand. Suppl.* 335 (1969).
4. S. Cohen, W. B. Long, W. J. Snape, in *International Review of Physiology, Gastrointestinal Physiology III*, vol 19 (R. K. Crane ed.) University Park Press, Baltimore (1979) p 128.
5. R. A. Gillis, J. D. Sousa, K. A. Hicks, A. W. Mangel, F. D. Pagani, B. L. Hamilton, T. Q. Harvey, D. G. Pace, R. K. Browne, W. P. Norman, *Am. J. Physiol.* 253: G531, (1987).
6. J. Christensen, M. J. Stiles, G. A. Rick, *Gastroenterology* 86: 706 (1984).
7. K. Fukai, H. Fukuda, *J. Physiol* 354: 89 (1984).
8. W. L. Hasler, S. Kurosawa, C. Owyang, *Am. J. Physiol.* 258: G404 (1990).
9. S. K. Sarna, *Dig. Dis. Sci.* 36: 827 (1991).
10. J. D. Hardcastle, C. V. Mann, *Gut* 9: 412 (1968).
11. R. C. Gill, K. R. Cote, K. L. Bowes, Y. J. Kingma, *Gut* 27: 293 (1986).
12. R. C. Gill, K. R. Cote, K. L. Bowes, Y. J. Kingma, *Gut* 27: 1006 (1986).
13. E. Daniel, K. L. Bowes, G. Duchon, in *Proceedings of the V International Symposium on Gastrointestinal Motilities.* (G. Vantrappen, ed.) Herentals Belgium Typoff Press (1975).
14. S. K. Sarna, B. L. Bardakjian, W. E. Waterfall, J. F. Lind, *Gastroenterology* 778: 1526 (1980).

15. J. C. Schang, M. Hermond, M. Herbet, M. Pilot, *Dig. Dis. Sci.* 31: 1331 (1986).
16. J. D. Huizinaga, E. E. Daniel, *Dig. Dis. Sci.* 31:865 (1986).
17. J. Christensen, S. Anuras, R. L. Hauser, *Gastroenterology* 66: 240 (1974).
18. T. K. Smith, J. B. Reed, K. M. Sanders, *Am. J. Physiol.* 252: C290 (1987).
19. L. Bueno, J. Fioramonti, *Clin. Res. Rev.* 1 Suppl 1: 91 (1981).
20. T. Y. El-Shakawy, *J. Physiol.* 342: 67 (1983).
21. H. L. Duthie, D. Kirk, *J. Physiol.* 283: 319 (1978).
22. N. G. Durdle, Y. J. Kingma, K. C. Bowes, M. M. Chambers, *Gastroenterology* 84: 375 (1983).
23. P. J. Sabourin, Y. J. Kingma, K. L. Bowes, *Am. J. Physiol.* 258: G484 (1990).
24. J. D. Huizinaga, E. Chow, N. E. Diamant, T. Y. El-Shakawy, *Am. J. Physiol.* 252: G136 (1987).
25. F. Narducci, G. Bassotti, M. Gaburri, A. Morelli, *Gut* 28: 17 (1987).
26. D. Kumar, N. S. Williams, D. Waldron, D. L. Wingate, *Gut* 30: 1007 (1989).
27. G. Bassotti, M. Gaburri, *Am. J. Physiol* 255: 660 (1988).
28. J. C. Schang, G. Devroede, M. Herbet, M. Hemond, M. Pilote, L. Devroede, *Dig. Dis. Sci.* 33: 614 (1988).
29. L. Bueno, J. Fioromanti, J. Frexinos, Y. Ruckebusch, *Hepatogastroenterology* 27: 281 (1980).
30. W. J. Snape, S. A. Matarazzo, S. Cohen, *Gastroenterology* 75: 373 (1978).
31. W. J. Snape, S. H. Wright, W. M. Battle, S. Cohen, *Gastroenterology* 77: 1235 (1979).
32. J. Christensen, *Am. J. Clin. Nutr.* 42: 1025 (1985).
33. L. F. Sillin, R. F. Condon, W. J. Shulte, J. H. Woods, P. O. Bass, V. W. L. Go, in *Gastrointestinal Motility in Health and Disease* (H. L. Duthie, ed.) MTP Press, Lancaster, England, (1978) p 361.
34. D. J. Holdstock, J. J. Misiewicz, *Gut* 11: 100 (1970).
35. M. J. Tansy, F. M. Kendall, *Am. J. Dig. Dis.* 18: 521 (1973).
36. A. M. Connell, S. T. D. McKelvey, *Proc. R. Soc. Med.* 63: 7 (1970).
37. T. Semba, *J. Med. Sci.* 2: 329 (1954).
38. D. J. Holdstock, J. J. Misiewicz, T. Smith, E. W. Rowlands, *Gut* 11: 91 (1970).
39. A. F. Hertz, A. Newton, *J. Physiol.* 47: 57 (1913).
40. M. L. Ramorino, C. Colagrande, *Am. J. Dig. Dis.* 9: 64 (1964).
41. Y. Ruckebusch, M. L. Grivel, M. J. Fargeas, *Physiol. Behav.* 6: 359 (1971).
42. A. Dahlgren, O. Selking, *Ups. J. Med. Sci.* 77: 167 (1972).

43. M. A. Sullivan, S. Cohen, W. J. Snape, *N. Engl. J. Med.* 298: 878 (1978).
44. E. A. Galapeaux, A. D. Templeton, *Am J. Physiol.* 119: 312 (1937).
45. S. Levinson, M. Blasker, T. R. Gibson, R. Morin, W. J. Snape, *Dig. Dis. Sci.* 30:33 (1985).
46. R. F. Harvey, A. E. Read, *Gut* 14: 983 (1973).
47. H. Meshkinpour, V. P. Dinoso, S. H. Lorber, *Gastroenterology* 66: 373 (1974).
48. S. K. Sarna, *Am. J. Physiol.* 250: G213 (1986).
49. J. Hinrichsen, A. C. Ivy, *Am. J. Physiol.* 96: 494 (1931).
50. M. J. Tansy, F. M. Kendall, J. J. Murphy, *Surg. Gynecol. Obstet.* 135: 404 (1972).
51. N. G. Kock, L. Hulten, L. Leandoer, *Scand. J. Gastroenterol.* 3: 163 (1968).
52. P. Kerlin, A. Zinsmeister, S. Phillips, *Gastroenterology* 84: 762 (1983).
53. S. R. Brown, P. A. Cann, N. W. Read, *Gut* 31: 454 (1990).
54. M. Camilleri, M. Neri, *Dig. Dis. Sci.* 34: 1777 (1989).
55. F. Narducci, W. J. Snape, W. M. Battle, R. L. Loudon, S. Cohen, *Dig. Dis. Sci.* 30: 40 (1985).
56. J. F. Erckenbrecht, B. H. Butsch, P. Enck, *Gastroenterology* 96: A141 (1989).
57. T. P. Almy, L. E. Hinkle, B. Berle, F. Kern, *Gastroenterology* 12: 437 (1949).
58. F. Narducci, G. Bassotti, M. Gaburri, A. Solinas, S. Fiorucci, A. Morelli, *Gastroenterology* 88: 1515 (1985).
59. A. Chauve, G. Devroede, E. Bastin, *Gastroenterology* 70: 336 (1976).
60. W. O. Kirwan, A. N. Smith, W. D. Mitchell, J. D. Falconer, M. A. Eastwood, *Gut* 16: 894 (1975).
61. C. A. Edwards, S. Brown, A. J. Baxter, J. J. Bannister, N. W. Read, *Gut* 30: 383 (1989).
62. R. C. Spiller, M. L. Brown, S. F. Phillips, *Gastroenterology* 91: 100 (1986).
63. T. Yajima, *J. Physiol.* 368: 667 (1985).
64. P. E. Squires, R. D. E. Rumsey, N. W. Read, *Gut* 31: A1170 (1990).
65. P. S. Kamath, S. F. Phillips, M. K. O'Connor, M. L. Brown, A. R. Zinsmeister, *Gut* 31: 443 (1990).
66. A. J. M. Brodribb, R. E. Condon, V. Cowles, J. L. De Cosse, *Gastroenterology* 77: 70 (1979).
67. C. Cherbut, Y. Ruckebusch, *Gastroenterol. Clin. Biol.* 8: 955 (1984).
68. W. O. Kirwan, A. N. Smith, A. A. McConnell, W. D. Mitchell, M. A. Eastwood, *Br. Med. J.* 4: 187 (1974).
69. J. Tomlin, N. W. Read, *Br. Med. J.* 297: 1175 (1988).
70. N. W. Read, C. A. Miles, D. Fisher, A. M. Holgate, N. D. Kime, M. A.

Mitchell, A. M. Reeve, T. B. Roche, M. Walker, *Gastroenterology* 79: 1276 (1980).

71. J. H. Cummings, D. J. A. Jenkins, H. S. Wiggins, *Gut* 17: 210 (1976).

72. A. Metcalfe, S. F. Phillips, A. R. Zinsmeister, R. L. MacCarty, R. W. Beart, B. G. Wolff, *Gastroenterology* 92: 40 (1987).

73. N. W. Read, M. N. Al-Janabi, A. M. Holgate, D. C. Barber, C. A. Edwards, *Gut* 27: 300 (1986).

74. B. Krevsky, L. S. Malmud, F. D'ercole, A. H. Maurer, R. S. Fisher, *Gastroenterology* 91: 102 (1986).

75. E. Moreno-Osset, G. Bazzocchi, S. Lo, B. Trombley, E. Ristow, S. N. Reddy, J. Villanueva-Mayer, J. W. Fain, J. Jing, I. Mena, W. J. Snape, *Gastroenterology* 96: 1265 (1989).

76. G. A. Spiller, M. C. Chernoff, R. A. Hill, J. E. Gates, J. J. Nassar, E. A. Shipley, *Am. J. Clin. Nutr.* 33: 754 (1980).

77. J. H. Cummings, M. J. Hill, D. J. A. Jenkins, J. R. Pearson, H. S. Wiggins, *Am. J. Clin. Nutr.* 29: 1468 (1976).

78. M. A. Eastwood, A. N. Smith, W. G. Brydon, J. Pritchard, *Gut* 19: 1144 (1978).

79. M. I. McBurney, P. J. Horvath, J. L. Jeraci, P. J. Van Soest, *Br. J. Nutr.* 53: 17 (1985).

80. J. Tomlin, N. W. Read, *Br. J. Nutr.* 60: 467 (1988).

81. J. P. Ryan, A. Bhojwani, *Am. J. Physiol.* 251: G46 (1986).

82. M. A. Kamm, M. J. G. Farthing, J. E. Lennard-Jones, *Gut* 30: 605 (1989).

83. S. S. Davis, J. G. Hardy, J. W. Fara, *Gut* 27: 886 (1986).

84. N. W. Read, K. Sugden, in *Therapeutic Drug Carrier Systems*, vol 4, Boca Raton: CRC Press, (1987) p 221.

85. K. P. Steed, P. J. Watts, L. Barrow, P. E. Blackshaw, C. D. Melia, M. C. Davies, C. G. Wilson, R. C. Spiller, *Gut* 31: A617 (1990).

86. J. G. Hardy, C. G. Wilson, E. Wood, *J. Pharm. Pharmacol.* 37: 874 (1985).

87. G. Parker, C. G. Wilson, J. G. Hardy, *J. Pharm. Pharmacol.* 40: 376 (1988).

88. J. G. Hardy, E. Wood, A. G. Clark, J. R. Reynolds, *Eur. J. Nucl. Med.* 11: 393 (1986).

89. J. G. Hardy, E. Wood, A. G. Clark, J. R. Reynolds, *Eur. J. Nucl. Med.* 12: 176 (1986).

90. G. D. Potter, S. M. Burlingame, *Am. J. Physiol.* 250: G221 (1986).

91. B. Braaten, J. L. Madara, M. Donowitz, *Am. J. Physiol.* 255: G72 (1988).

92. J. C. Debongie, S. F. Phillips, *Gastroenterology* 74: 698 (1978).

93. R. Palem, N. Vidon, J. J. Bernier, *Dig. Dis. Sci.* 26: 929 (1981).

94. J. H. Sellin, R. De Soignie, *Gastroenterology* 93: 441 (1987).

95. K. A. Hubel, K. Renquist, S. Shirazi, *Gastroenterology* 92: 501 (1987).
96. R. Levitan, J. S. Fordtran, B. A. Burrows, F. J. Ingelfinger, *J. Clin. Invest.* 41: 1754 (1962).
97. G. J. Devroede, S. F. Phillips, C. F. Code, J. F. Lind, *Can. J. Physiol. Pharmacol.* 49: 1023 (1971).
98. N. I. McNeil, in *Intestinal Absorption and Secretion*, Falk Symposium 36 (E. Skadhauge, K. Heintze eds.) Lancaster, England, MTP Press (1984).
99. E. S. Foster, M. E. Budinger, J. P. Hayslett, H. J. Binder, *J. Clin. Invest.* 77: 228 (1986).
100. H. J. Binder, E. S. Foster, M. E. Budinger, J. P. Hayslett, *Gastroenterology* 93: 449 (1987).
101. R. Lübke, K. Haag, E. Berger, H. Knauf, W. Gerok, *Am. J. Physiol.* 251: G132 (1986).
102. G. I. Sandle, N. K. Wills, W. Alles, H. J. Binder, *Gut* 27: 999 (1986).
103. G. Lonnerholm, O. Selkin, P. J. Wistrand, *Gastroenterology* 88: 1151 (1985).
104. E. S. Foster, J. P. Hayslett, H. J. Binder, *Am. J. Physiol.* 246: G611 (1984).
105. W. Clauss, H. Hornicke, *Compr. Biochem. Physiol.* 79A: 267 (1984).
106. S. K. Sullivan, P. L. Smith, *Am. J. Physiol.* 250: G474 (1986).
107. U. Kück-Bierre, W. Von Englehardt, *Gut* 31: 64 (1990).
108. P. L. Smith, R. D. McCabe, *Am. J. Physiol.* 247: 445 (1984).
109. W. V. Engelhardt, U. Kuck, M. Krause, *Pflugers. Arch.* 407: 625 (1986).
110. R. J. Bridges, M. Rack, W. Rummel, J. Schriener, *J. Physiol.* 376: 531 (1986).
111. A. Kuwahara, H. J. Radowicz-Cooke, *J. Physiol.* 395: 271 (1988).
112. C. A. Edwards, J. Bowen, W. G. Brydon, M. A. Eastwood, *Br. J. Nutr.* 68: 473 (1992).
113. D. Bleakman, R. J. Naftalin, *Am. J. Physiol.* 258: G377 (1990).
114. M. Shiga, T. Muraoka, T. Hirsawa, S. Awazu, *J. Pharm. Pharmacol.* 37: 446 (1985).
115. W. E. W. Roedigger, *Gastroenterology* 83: 424 (1982).
116. J. H. Cummings, W. J. Branch, E. W. Pomare, C. P. E. Naylor, G. T. MacFarlane, *Gut* 28: 1221 (1987).
117. R. A. Argenzio, M. Southworth, J. E. Lowe, C. E. Stevens, *Am. J. Physiol.* 233: E469 (1977).
118. H. Ruppin, S. Bar-Meir, K. H. Soergel, C. M. Wood, M. G. Schmitt, *Gastroenterology* 78: 1500 (1980).
119. N. I. McNeil, K. L. E. Ling, J. Wager, *Gut* 28: 707 (1987).
120. Y. Suzuki, K. Kaneto, *Am. J. Physiol.* 253: G155 (1987).
121. R. A. M. Argenzio, M. Southworth, J. E. Lowe, C. E. Stevens, *Am. J. Physiol.* 233: E469 (1977).

122. H. J. Binder, P. Mehta, *Gastroenterology* 96: 989 (1989).
123. M. Hatch, *Am. J. Physiol.* 253: G171 (1987).
124. B. S. Ramakrishna, S. H. Nance, I. C. Roberts-Thomson, W. E. W. Roediger, *Digestion* 45: 93 (1990).
125. R. W. McNealy, J. D. Willems, *Surg. Gynecol. Obstet.* 49: 794 (1929).
126. F. S. Curry, J. A. Bargen, *Surg. Gynecol. Obstet* 60: 667 (1935).
127. L. Slade, R. Bishop, J. G. Morris, D. W. Robinson, *Br. Vet. J.* 127: 11 (1971).
128. S. C. Kathpalia, M. J. Favus, F. L. Coe, *J. Clin. Invest* 63: 805 (1984).
129. C. O. Billich, R. Levitan, *J. Clin. Invest.* 48: 1336 (1969).
130. J. S. Fordtran, F. C. Restor, M. F. Ewton, N. Soter, J. Kinney, *J. Clin. Invest* 44: 1935 (1965).
131. L. S. Schanker, *J. Pharmacol.* 126: 283 (1959).
132. G. Rechkemmer, M. Wahl, H. Kuschinski, W. V. Engelhardt, *Pflugers Arch.* 407: 33 (1986).
133. G. W. Smith, P. M. Wiggins, S. P. Lee, C. Tasman-Jones, *Clin. Sci.* 70: 271 (1986).
134. J. R. Vercellotti, A. A. Salyers, W. S. Bullard, T. D. Wilkins, *Can. J. Biochem.* 55: 1190 (1977).
135. C. A. Edwards, N. W. Read, In *Dietary Fibre Perspectives*, vol. 2 (A. R. Leeds, V. J. Burley, eds.) London, John Libbey, (1990) p 52.
136. G. V. Vahouney, R. Tombes, M. M. Cassidy, D. Kritchevsky. L. L. Gallo, *Proc. Soc. Exp. Biol. Med.* 166: 12 (1981).
137. M. A. Eastwood, D. Hamilton, *Biochem. Biophys. Acta* 152: 165 (1968).
138. A. Wise, A. K. Mallett, I. R. Rowland, *Toxicology* 38: 241 (1986).
139. I. Rowland, A. K. Mallett, In *Dietary Fibre: Chemistry, Physiology and Health Effects* (D. Kritchevsky, C. Bonfield, J. W. Anderson, eds.) New York, Plenum Press (1990) p 195.
140. G. Pye, J. Crompton, D. F. Evans, A. G. Clarke, J. D. Hardcastle, *Gut* 28: A1328 (1987).
141. J. R. Lipton, D. M. Coder, L. R. Jacobs, *J. Nutr.* 118: 840 (1988).
142. S. J. Meldrum, B. J. Watson, H. C. Riddle, R. L. Bown, G. E. Sladen, *Br. Med. J.* 2: 104 (1972).
143. R. C. Bown, J. A. Gibson, G. E. Sladen, B. Hicks, A. W. Dawson, *Gut* 15: 999 (1974).
144. C. A. Edwards, J. Bowen, M. A. Eastwood, in *Dietary Fibre: Chemical and Biological Aspects* (D. A. T. Southgate, K. Waldron, I. T. Johnson, eds.) Cambridge, Royal Society of Chemistry (1990) p 273.
145. T. Sakata, *Br. J. Nutr.* 58: 95 (1987).
146. F. V. Mortensen, H. Nielsen, M. J. Mulvany, I. Hessov, *Gut* 31: 1391 (1990).
147. P. R. Kvietys, D. N. Granger, *Fed. Proc.* 41: 2100 (1982).

148. A. S. Grandison, J. Yates, R. Shields, *Gut* 22: 223 (1981).
149. S. Fasth, L. Hulten, O. Lungren, S. Nordgren, *Acta Physiol. Scand.* 101: 98 (1977).
150. P. R. Kvietys, D. N. Granger, *Gastroenterology* 80: 962 (1981).
151. J. D. Fondacaro, in *Physiology of the Intestinal Circulation* (A. P. Shepherd, D. N. Granger, eds.) New York, Raven Press, (1984) p 107.

2

Drug Metabolism in the Colon Wall and Lumen

Johann W. Faigle*

Ciba-Geigy Limited, Basel, Switzerland

I. INTRODUCTION

Synthetic drugs are normally lipophilic, at least to some extent, and can pass through the lipid membranes of the body by passive diffusion. However, strongly lipophilic molecules cannot be easily excreted from the mammalian organism unless they are converted into hydrophilic metabolites. As a rule, metabolism of drugs and other xenobiotics is enzyme-catalyzed and proceeds in two distinct phases (1, 2).

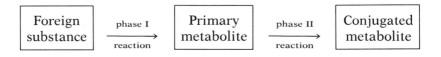

In phase I, a functional group is introduced into the molecule by an oxidative, reductive, or hydrolytic reaction. This group is then

---

* Retired

linked, or conjugated, with an endogenous polar substituent in phase II. Introduction of a hydroxyl group and conjugation with glucuronic acid may serve as a typical example for the two consecutive phases of metabolism. Although the primary metabolites often exert pharmacologic or toxicologic effects, the conjugates formed as end products are generally inactive. The enzymes catalyzing phase I and II reactions are mainly located in the liver, but are also present in the intestine, kidney, lung, and other tissues.

The gastrointestinal tract is the major port of entry for drugs into the human body. In contrast to parenteral routes of administration however, the oral route may be subject to first-pass metabolism by enzymes present in the wall or lumen of the alimentary canal or in the liver. A hepatic or intestinal first-pass metabolism may reduce the systemic availability of an active substance and may thereby interfere with the therapeutic result. However, prodrugs are often specially designed to become activated through first-pass metabolism.

Metabolic reactions in the liver and small intestine are well documented (1, 3–8). However, only fragmentary or sparse information is available on drug metabolism in the colon. First, it is technically difficult to examine the fate of a substance in the small and large intestine particularly in humans. Second, most of the classic drugs intended for oral use are rapidly absorbed and therefore do not reach the colon to any large extent. Until recently, it was generally believed that the colonic mucosa efficiently only absorbs water and electrolytes from the lumen, but does not take up drugs and other xenobiotics.

With the advent of oral formulations allowing controlled release of drugs over 12 hr or longer, the need for better knowledge of metabolism and presystemic clearance in the colonic lumen and wall became obvious. A few studies with individual drugs have been done. What is known in addition stems from investigations on chemical carcinogenesis, which involves metabolic activation in the gut wall. Some information is also available from studies on the effects of the intestinal microflora on foreign compounds.

This chapter will review, summarize, and assess the published information on this topic. In the first part, some general physiological factors influencing or governing drug metabolism in the gut will be described. The next part will deal with reactions mediated by the host enzymes of the intestinal mucosa and by the microbial

enzymes of the gut flora. The pharmacologic and toxicologic importance of biotransformation in the colon will also be discussed. This
chapter will focus on observations made in the human organism,
although results from animal models will also be considered if appropriate.

II. PHYSIOLOGICAL FACTORS INFLUENCING DRUG METABOLISM IN THE GUT

The gut has a series of muscular coats, lined by a single layer of
mucosal cells. These cells, the enterocytes, are tightly connected
and form the main barrier between the gut lumen and the circulating blood in the intestinal tissue. However, most synthetic drugs
intended for oral use are sufficiently lipophilic to penetrate the cell
membrane by passive diffusion and to move inside the cell. The
metabolizing enzymes of the gut wall are exclusively found in the
enterocytes, where they act as a first line of defense against lipophilic xenobiotics (3, 4).

Metabolic and absorptive functions of the intestinal mucosa,
therefore, are closely associated. Both will change as the total surface of the mucosa changes from one gut section to another. Drug
molecules may also be attacked by enzymes present in the intestinal microflora. The metabolic capacity of the flora depends on
the number of micro-organisms, which again changes along the
axis of the gastrointestinal tract (7, 9).

To understand better the metabolic capacities of the colon and
of other sections of the human gut, some physiological particulars
are summarized in Table 1. The small intestine has a mucosal surface of about 120 m^2, whereas that of the large intestine is only in
the range of 0.3 m^2 (10, 11). The enormous total area of duodenum,
jejunum, and ileum is provided by concentric folds of the mucosa
and by tiny projections of the mucosal surface, called macrovilli.
A brush border consisting of microvilli covers the absorbing cells
and enlarges the mucosal surface of the small intestine even more
(4500 m^2, Table 1). The colon does not contain any folds or villi and
is shorter than the small intestine.

The cells at the tips of the macrovilli contain larger amounts of
drug-metabolizing enzymes than other enterocytes (4). This may
explain why the specific activity of an enzyme, calculated per milli-

Table 1 Physiological Factors Influencing Drug Absorption and Metabolism in Various Sections of Human Gastrointestinal Tract (Approximate or Average Values)

Section	Length (m)	Surface $(m^2)^a$	pH	Micro-organisms $(count/g)^b$	Transit time (h)
Stomach	0.2	0.1	1.5	$\leq 10^2$	variable
Small intestine	7	120^c			5
Duodenum	0.3	0.1	6.9	$\leq 10^2$	
Jejunum	3	60	6.9	$\leq 10^5$	
Ileum	4	60	7.6	$\leq 10^7$	
Large intestine	1.5	0.3	8.0	$\geq 10^{11}$	≤ 48

Sources: Refs. 7, 9–15, 53, 54.
[a] Absorptive area of the mucosa
[b] Number of micro-organisms per gram of gastrointestinal content
[c] The large absorptive area (120 m²) of the small intestine is attributable to submucosal folds (plicae) and mucosal projections (macrovilli). This area is even larger (4500 m²), when the brushborder (microvilli) covering the absorbing cells is taken into account. The mucosa of the colon has no folds and no villi.

gram protein, is often lower in the colonic than in the small intestinal mucosa. The total metabolic capacity of the large intestine must be inferior to the small intestine, if one considers the differences in length and inner surface.

Because of the low gastric pH there are virtually no micro-organisms in the stomach of healthy humans. The number of organisms increases gradually along the small intestine, but it rises by several orders of magnitude beyond the ileocecal valve (Table 1). The metabolizing force represented by the human microflora is concentrated in the large intestine (7, 9, 12–15).

Drugs may also be attacked by host enzymes secreted into the lumen of the gut. It is well known that hydrolyzing enzymes such as amylases, peptidases, and esterases are shed into the lumen of the small intestine. Ester or amide bonds present in foreign substances may therefore be cleaved. However, such reactions will be limited to the upper intestine, since the gut flora will degrade these enzymes in the lower bowel (9, 10).

From these general considerations one can already conclude that a drug reaching the large intestine will be metabolized by microbial enzymes in the first place. If a drug molecule is taken up by the enterocytes of the colonic mucosa, it may also be attacked there, although the total metabolic capacity at this site of the gut is lower than that in the small intestine. Host enzymes secreted into the lumen of the gut will not significantly contribute to colonic bio-transformation.

Another general factor to be considered is the speed at which the contents of the gastrointestinal tract move through the various sections. The residence time of solids in the stomach is variable, but liquid matter and small particles of ≤ 2 mm diameter are con-tinuously emptied through the pylorus. The transit time in the small intestine is up to 5 hr while that in the large intestine ranges from a few hours to 2 days (Table 1) (9, 12, 16).

An orally ingested drug that is readily dissolved and rapidly absorbed will not reach the lower bowel. If one assumes a first-order absorption process controlled by a rate constant k_A, a sub-stance with $k_A = 5 \, hr^{-1}$ will be almost completely absorbed within 30 min of administration. Even if absorption proceeds much more slowly, for example, with $k_A = 0.5 \, hr^{-1}$, it would approach 90% after 5 hr (16).

In principle, there are four ways by which a xenobiotic comes in contact with the metabolizing enzymes in the wall or lumen of the colon: (1) a drug substance that is sparingly soluble or strongly polar may escape absorption in the stomach and small intestine following oral administration; (2) polar metabolites of xenobiotics formed in the liver may be excreted into the gut lumen by the biliary route; (3) therapeutic systems not retained in the stomach and designed to release a drug during more than 5 hr will deliver part of the dose into the colon; (4) therapeutic systems designed to target a drug to the colon will, in an ideal case, deliver the entire dose to this part of the gut.

In this way, colonic drug metabolism may often become impor-tant. In the following pages the metabolic reactions taking place and the enzymes catalyzing them will be systematically compiled.

III. DRUG METABOLISM BY ENZYMES IN THE COLON WALL

A. Enzymes in the Colonic Mucosa

The individual cells representing the colonic mucosa have similar metabolic functions to the cells in the small intestine, although the

Table 2 Drug-Metabolizing Enzymes in the Human Colonic Mucosa:
Enzymes Catalyzing Phase I Reactions

Enzyme	Location in cell fraction	Inducibility by xenobiotics	Metabolic reaction catalyzed
Cytochrome P-450	Microsomes	Marked	Oxidation of C-, N-, or S-atoms
Epoxide hydrolase	Microsomes and cytosol[a]	Marked	Cleavage of 1,2-epoxides
Esterases, amidases	Microsomes and cytosol[a]	Slight	Cleavage of esters or amides of carboxylic acids
Glucuronidase	No data available	Slight	Cleavage of beta-glucuronides of alcohols and phenols

Sources: Refs. 6, 20–26.
[a] Enzyme activity in microsomes is several times that in cytosol.

two gut sections differ in their surface and their architecture, as
explained above. The enterocytes at both sites contain various enzymes capable of catalyzing a wide range of metabolic reactions.
Lipophilic xenobiotics are particularly susceptible to metabolism,
since they readily gain access to the enzymes inside the enterocytes.
The resulting polar metabolites may enter the bloodstream or be
secreted back into the gut lumen (10, 11, 17–19).

The major enzymes in phase I and phase II metabolism in the
colon wall of humans are summarized in Tables 2 and 3. Details
on substrates and chemical reactions will be covered in section
III.B. Some enzymes are embedded in the smooth endoplasmic
reticulum of the enterocytes while some others are present in dissolved form in the cytoplasm. When subcellular fractions are prepared by centrifugation, the former enzymes are mostly in the microsomal pellet, and the latter ones in the supernatant (17).

Cytochrome P-450 is a microsomal enzyme catalyzing oxidative
reactions (Table 2). Epoxide hydrolase is also largely present in the
microsomes. The two enzymes are functionally interlinked, since
reactive epoxides formed by P-450 in the gut wall are detoxified

Table 3 Drug-Metabolizing Enzymes in the Human Colonic Mucosa: Enzymes Catalyzing Phase II Reactions

Enzyme	Location in cell fraction	Inducibility by xenobiotics	Metabolic reaction catalyzed
Glucuronyl transferase	Microsomes, mitochondria and nuclei	Marked	Glucuronidation of alcohols, phenols, and amines
Sulfotransferase	Cytosol	None	Sulfation of phenols and steroids
Acetyl transferase	Cytosol	No data available	Acetylation of amines
Glutathione S-transferase	Cytosol	Slight	Glutathione transfer to electrophilic drug metabolites

Source: Refs. 3, 4, 7, 8, 24, 27–41.

by the hydrolase. From animal experiments it is known that both enzymes are inducible, for instance, by phenobarbital and 3-methylcholanthrene. Only one isoenzyme of cytochrome P-450 is detectable in the human colon, whereas several isoforms are present in the liver and small intestine. This implies that the spectrum of reactions catalyzed by P-450 may be less diverse in the colon than at the other sites (6, 20–25).

The remaining enzymes listed in Table 2 can hydrolyze ester, amide, or glycoside bonds in foreign substances. Hydrolysis in the gut wall may inactivate a peptide drug, but the same mechanism can be deliberately used to release the active moiety of an ester prodrug. Certain drug glucuronides formed in the liver and excreted in the bile may be cleaved by glucuronidases in the enterocytes. The free drug molecules may then enter the bloodstream again. However, this enterohepatic circulation is more likely to be mediated by the gut flora than by the gut wall (4, 23, 26).

Hepatic phase I metabolism comprises three types of reactions: oxidation, reduction, and hydrolysis (see section I). Oxidative and hydrolytic enzymes are definitively present in the colonic mucosa, but there is no information available on reductive ones such as nitro- or carbonyl reductases. Enzymatic reduction of drugs in the gut wall seems not to be important as a whole.

The many transferases catalyzing phase II reactions in the colonic enterocytes (Table 3) are essentially the same as in hepatocytes. These transferases are cytosolic enzymes, except for glucuronyl transferase, which is present in the microsomal, mitochondrial, and nuclear fractions of the colonic cells. Only the latter enzyme is reported to be readily inducible. It should be kept in mind that the conjugates resulting from phase II metabolism are normally without biological activities (4, 7, 8, 27, 28).

Phenolic drugs and metabolites are the major substrates for both glucuronyl transferase and sulfotransferase (Table 3). The two enzymes often compete for the same substrate. However, glucuronidation is a high-capacity, low-affinity reaction, and sulfation is a low-capacity, high-affinity reaction. Therefore, sulfation normally predominates at low substate concentrations in the colon and glucuronidation at high ones (3, 29–34).

Acetylation of xenobiotic amines in the human colon occurs quite often, but shows a high degree of interindividual variation. The larger part of this variation is resulting from a genetic polymorphism of the intestinal acetyl transferase. The hepatic transferase displays the same type of polymorphism, giving rise to the phenotypes known as "fast" and "slow" acetylators (3, 4, 8, 24, 35).

Conjugation with glutathione is an effective mechanism by which electrophilic, potentially hazardous intermediates of phase I metabolism can be disposed of. This defensive mechanism is also active in the large intestine of humans, although the colonic enterocytes contain one acidic isoform of glutathione S-transferase only. Other organs such as liver and small intestine also contain basic and neutral isoenzymes (36–41).

In conclusion, the drug-metabolizing enzymes present in the human colon wall can catalyze a variety of oxidative, hydrolytic, and conjugative reactions. The spectrum of enzymes in the enterocytes is not as diversified as that in the hepatocytes, however.

B. Metabolic Reactions in the Colonic Mucosa

From the scattered data available on colonic drug metabolism it is difficult to get an understanding of the types and mechanisms of the underlying chemical reactions. Therefore, an attempt was made to compile and summarize the essential reactions in a systematic way. The results are given in Tables 4–6 (1, 2, 42).

Oxidative reactions may take place at carbon atoms or at hetero atoms in drug molecules (Table 4). Hydroxylation of aromatic and

Table 4 Oxidative Drug Metabolism in the Intestinal Mucosa

Oxidation of carbon atoms		
Aromatic rings	$R\!-\!C_6H_5$	$\rightarrow R\!-\!C_6H_4\!-\!OH$
Aliphatic chains	$\begin{array}{c}R^1\\ \backslash\\ CH_2\\ /\\ R^2\end{array}$	$\rightarrow\quad\begin{array}{c}R^1\\ \backslash\\ CH\!-\!OH\\ /\\ R^2\end{array}$
Ethers	$R\!-\!O\!-\!CH_3$	$\rightarrow [R\!-\!O\!-\!CH_2OH] \xrightarrow{-HCHO} R\!-\!OH$
Amines, *prim*	$R\!-\!CH_2\!-\!NH_2$	$\rightarrow [R\!-\!CH(OH)\!-\!NH_2] \xrightarrow{-NH_2} R\!-\!CHO$
Amines, *sec, tert*	$\begin{array}{c}R^1\\ \backslash\\ N\!-\!CH_3\\ /\\ R_2\end{array}$	$\rightarrow \left[\begin{array}{c}R^1\\ \backslash\\ N\!-\!CH_2OH\\ /\\ R^2\end{array}\right] \xrightarrow{-HCHO} \begin{array}{c}R^1\\ \backslash\\ NH\\ /\\ R^2\end{array}$
Oxidation of hetero atoms		
Amines, *prim, sec*	$\begin{array}{c}R^1\\ \backslash\\ N\!-\!H\\ /\\ R^2\end{array} \rightarrow$	$\begin{array}{c}R^1\\ \backslash\\ N\!-\!OH\\ /\\ R^2\end{array}$
Amines, *tert*	$\begin{array}{c}R^1\\ \backslash\\ R^2\!-\!N\\ /\\ R^3\end{array} \rightarrow$	$\begin{array}{c}R^1\\ \backslash\\ R^2\!-\!N\rightarrow O\\ /\\ R^3\end{array}$
Sulfides (thioethers)	$\begin{array}{c}R^1\\ \backslash\\ S\\ /\\ R^2\end{array} \rightarrow \begin{array}{c}R^1\\ \backslash\\ S\rightarrow O\\ /\\ R^2\end{array} \rightarrow$	$\begin{array}{c}R^1\quad O\\ \backslash\; \nearrow\\ S\\ /\; \searrow\\ R^2\quad O\end{array}$

Sources: Refs. 1, 2, 42.

aliphatic carbon yields phenols and alcohols. Formation of phenols proceeds normally via arene oxides:

$$\underset{\diagup}{\overset{H}{\diagdown}}C=C\underset{\diagdown}{\overset{H}{\diagup}} \rightarrow \underset{\diagup}{\overset{H}{\diagdown}}C\underset{\diagdown}{\overset{O}{\diagup\diagdown}}C\underset{\diagdown}{\overset{H}{\diagup}} \rightarrow \underset{\diagup}{\overset{H}{\diagdown}}C=C\underset{\diagdown}{\overset{OH}{\diagup}}$$

If hydroxylation takes place at alkyl groups next to an O- or N-atom in ethers or amines, dealkylation, or deamination occurs. The intermediates of such reactions are instable and split off an aldehyde or ammonia. Oxidation at hetero atoms produces hydroxylamines and N-oxides from amines and sulfoxides and sulfones from thioethers.

The hydrolytic reactions shown in Table 5 are straightforward and need little comment. It may be worth mentioning that the 1,2-diols arising from the cleavage of epoxides have *trans* and not *cis* configuration.

All conjugative reactions need endogenous ligands (Table 6). The glucuronyl moiety is supplied by uridine diphosphate glucuronic acid (UDPGA), and sulfate is provided by 3'-phosphoadenosine-5'-

Table 5   Hydrolytic Drug Metabolism in the Intestinal Mucosa

Hydrolysis of C—O and C—N bonds		
Carboxylic acid esters	R^1—CO—OR^2	$\rightarrow R^1$—COOH + R^2—OH
Carboxylic acid amides	R^1—CO—$N\diagup^{R^2}_{\diagdown R^3}$	$\rightarrow R^1$—COOH + $HN\diagup^{R^2}_{\diagdown R^3}$
Epoxides, arene oxides	H—$\underset{R^1}{\overset{O}{C\diagup\diagdown}}$——$\underset{R^2}{C}$—H $\rightarrow$	H—$\underset{R^1}{\overset{HO}{C}}$—$\underset{R^2}{\overset{H}{C}}$—OH
Glucuronides	R—CH_2O—Gluc[a]	$\rightarrow$ R—CH_2OH + GA[b]

Sources: Refs. 1, 2, 42.
[a] Ethereal O-glucuronide of an alcohol as an example.
[b] Glucuronic acid (in free form).

Table 6   Conjugative Drug Metabolism in the Intestinal Mucosa

Conjugation of aryl (Ar)[a] derivatives with:		
Glucuronic acid (GA)	Ar—OH + GA	→ Ar—O—Gluc
Sulfuric acid	Ar—OH + H_2SO_4	→ Ar—O—SO_3H
Glutathion (GSH)	[Ar—X][b] + GSH	→ Ar—SG
Acetic acid	Ar—NH_2 + CH_3COOH	→ Ar—NH—$COCH_3$

Sources: Refs. 1, 2, 42.
[a] aryl (rather than alkyl) derivatives are the preferred substrates for conjugation
[b] Electrophilic intermediate of Phase 1 metabolism

phosphosulfate (PAPS). Glutathione is a tripeptide (glutamyl cysteyl glycine), which is present in free form in mammalian tissues. The ligand needed for acetylation stems from acetyl coenzyme A. The glucuronides, sulfates, and acetyl derivatives are end products of metabolism. The glutathione conjugates are mostly further metabolized to acetyl cysteyl derivatives, known as mercapturic acids.

Selected examples of drugs undergoing biotransformation in the gut wall are listed in Table 7. Metabolism was shown in humans for most of these drugs, but sometimes investigations were limited to the small intestine. However, if a drug reaches the colonic mucosa, the same reactions will occur there as well, because the drug-metabolizing enzymes are qualitatively similar at both sites. The examples suggest that hydrolytic reactions and conjugations are more important than oxidative reactions (3, 5, 7, 17–19, 23, 38, 43).

C. Differences Between Metabolism in Colon Wall and Liver

Some qualitative differences between the enzymatic spectrum in the liver and in the intestinal wall have been addressed. To summarize, the colonic enterocytes contain less isoforms of certain enzymes than the hepatocytes and reductive enzymes are largely missing in the intestinal wall. However, the quantitative differences between liver and gut wall metabolism seem to be even more important than the qualitative ones. The content of microsomal protein is much lower in the enterocytes than in the hepatocytes.

Table 7 Examples of Drugs that are Substrates for Enzymes in the
Intestinal Mucosa

Substrate	Reaction	Enzyme
Acetaminophen	Sulfation of OH-group	Sulfotransferase
	Glucuronidation of OH-group	Glucuronyl transferase
Acetyl salicylic acid	Hydrolysis of ester bond	Esterase
Alacepril	Hydrolysis of amide bond	Amidase
5-Amino salicylic acid	Acetylation of NH_2–group	Acetyl transferase
Chlorpromazine	Oxidation of S-atom	Cytochrome P-450
Ethinylestradiol	Sulfation of OH-group	Sulfotransferase
	Hydroxylation of C-atom	Cytochrome P-450
Flurazepam	Oxidative N-deethylation	Cytochrome P-450
Isoniazid	Acetylation of NH_2-group	Acetyl transferase
Isoprenaline	Sulfation of OH-group	Sulfotransferase
Isosorbide dinitrate	Conjugation with glutathione	Glutathione transferase
Lorazepam	Glucuronidation of OH-group	Glucuronyl transferase
Morphine	Glucuronidation of OH-group	Glucuronyl transferase
Pivampicillin	Hydrolysis of ester bond	Esterase
Sulfonamides	Acetylation of NH_2-group	Acetyl transferase

Sources: Refs. 3, 5, 7, 17–19, 23, 38, 43

Human small or large intestinal mucosa contains about 3.5 mg microsomal protein/g tissue, compared to 30 mg/g in the liver. There is not such a big difference in the content of cytosolic protein. The values are about 30 and 50 mg/g (24, 35). These figures suggest already that the metabolic capacity of the gut wall must be smaller than that of the liver, particularly for those enzymes present in the microsomes (Tables 2, 3). The specific activity of an enzyme, calculated per milligram protein, is also often lower in enterocytes than in hepatocytes.

If one takes these differences and the human organ weights into account, the gut to liver ratio was judged to be as low as 1:100 for the total activity of microsomal enzymes, such as cytochrome P-450 and glucuronyl transferase. The difference is less pronounced for cytosolic enzymes. Glutathione S-transferase with a gut to liver ratio of about 1:10 may serve as an example (44). These figures stand for the small intestine only. Colon and small intestine resemble each other regarding the spectrum of drug metabolizing enzymes (24, 35). However, they differ in the mass of enterocytes since the mucosal surface is much larger in the upper part of the gut (Table 1). Therefore, the total metabolic capacity of the colonic mucosa is by far inferior to the liver.

When a conventional drug formulation is administered by the oral route, the local concentrations in the intestine may become high. Because of that, saturation of drug-metabolizing enzymes will more often be encountered in the gut wall than in the liver also because of the low enzyme activity at the former site (19). Saturation of enzymes in the gut wall is less probably to occur with modern delivery systems, which release a drug at low concentrations during several hours. Nevertheless, first-pass metabolism in the gut wall may become important in all parts of the bowel including the colon. In general, conjugative and hydrolytic reactions will be more important than oxidative ones.

IV. DRUG METABOLISM BY ENZYMES OF THE INTESTINAL FLORA

A. Enzymes in Intestinal Micro-Organisms

It has been claimed there are more cells within the intestinal lumen than elsewhere in the human body. The lower gut contains about 10^{11} micro-organisms/g (Table 1), which represent more than 400 species. *Bacteroides* and bifidobacteria are predominant while *Clostridia*, enterobacteria, enterococci, and lactobacilli are quantitatively less important. This complex system comprises both strict anaerobes and facultative anaerobes. However, the colonic envi-

ronment in which the flora exists is oxygen free so that even facultative anaerobes metabolize anaerobically (7, 9, 13, 45–47).

The intestinal flora contain a wide spectrum of enzymes that
catalyze various metabolic reactions. However, there is a basic difference between drug metabolism in the liver and in the gut lumen.
The liver tends to oxidize and the flora to reduce xenobiotics because of the absence of oxygen in the gut. A similar type of comple-
mentary action takes place with conjugates of xenobiotics: Hepatic
enzymes mediate the synthesis of conjugates whereas microbial
enzymes often hydrolyze such conjugates.

Reduction and hydrolysis are the predominating processes in
the lumen of the colon. The enzymes are given in Tables 8 and 9,
with the species of micro-organisms known to contain such enzyme
activities. The list of micro-organisms is not meant to be complete,
but should illustrate how complex the assignment of enzymes is.
So there is little or no information available on the activity, specificity, and molecular biology of individual enzymes (7, 9, 12–15,
45, 46, 48).

Reductive enzymes attack functional groups by either introduc-
ing hydrogen or removing oxygen atoms. Most of these reactions
may also occur in the liver, at least to some degree. However, hy-

Table 8 Drug-Metabolizing Enzymes in the Human Intestinal Flora:
Enzymes Catalyzing Reductive Reactions

Enzyme	Sample or micro-organism investigated	Metabolic reaction catalyzed
Nitroreductase	Intestinal content (*E. coli*, bacteroides)	Reduction of aromatic and heterocyclic nitro compounds
Azoreductase	Feces (clostridia, lactobacilli)	Reductive cleavage of azo compounds
N-Oxide reductase, sulfoxide reductase	Feces; intestinal content (*E. coli*)	Removal of oxygen from N-oxides and sulfoxides
Hydrogenase	Feces (clostridia, lactobacilli)	Reduction of carbonyl groups and aliphatic double bonds

Sources: Refs. 7, 9, 45, 46

Table 9 Drug-Metabolizing Enzymes in the Human Intestinal Flora:
Enzymes Catalyzing Hydrolytic Reactions

Enzyme	Sample or micro-organism investigated	Metabolic reaction catalyzed
Esterases and amidases	Feces; intestinal content (*E. coli, P. vulgaris, B. subtilis, B. mycoides*)	Cleavage of esters or amides of carboxylic acids
Glucosidase	Intestinal content (clostridia, eubacteria, *Strep. faecalis*)	Cleavage of beta-glycosides of alcohols and phenols
Glucuronidase	Feces (*E. coli, A. aerogenes*)	Cleavage of beta-glucuronides of alcohols and phenols
Sulfatase	Feces (enterobacteria, clostridia, streptococci)	Cleavage of O-sulfates and sulfamates

Sources: Refs. 7, 9, 12, 14, 15, 45, 46, 48

drogenation of C=C double bonds is presumably catalyzed by microbial enzymes only (Table 8). Hydrolytic enzymes of the flora cleave all kinds of ester, amide, and glycoside bonds and play an important role in the enterohepatic circulation of drugs (Table 9). For more details on substrates and reactions see section IVB.

Unlike the enzymes in liver and gut wall, those in the gut flora cannot be considered as a first line of defenses against lipophilic and potentially toxic xenobiotics. Most of these xenobiotics are generally already absorbed in the upper part of the gastrointestinal tract and do not reach the colon to any great extent. However, if a xenobiotic comes into contact with the flora, metabolism quite often results in activation instead of inactivation. This is particularly true for hydrophilic conjugates, when ingested or secreted through the bile. However, it may also happen with other poorly absorbable foreign substances.

The activities of the microbial enzymes in the human gut are known to be influenced by several factors, such as age and diet. Enzyme induction upon prolonged exposure to chemical substances does also occur, although the enzymes in the flora seem to adapt less readily than those in the host tissues. The microbial enzymes for which induction has specifically been reported include reductive ones, such as azo- and nitroreductase and hydrolytic ones, such as glucosidase, glucuronidase, and sulfatase (9, 14, 15, 45).

B. Metabolic Reactions in Intestinal Micro-organisms

Some chemical aspects of reductive drug metabolism in the flora are summarized in Table 10 (1, 2, 42). Nitro groups in aromatic or heterocyclic molecules are converted to primary amines, hydroxyl amines being formed as intermediates. Reductive cleavage of aromatic azo compounds results in two primary amines. Removal of oxygen from an N-oxide or a sulfoxide yields the corresponding tertiary amine or the thioether. Alcohols are formed by the reduc-

Table 10 Reductive Drug Metabolism by the Intestinal Microflora

	Reduction of carbon or hetero atoms		
Nitro groups	$Ar—NO_2$	$\to Ar—NHOH \to Ar—NH_2$	
Azo groups	$Ar^1—N{=}N—Ar^2$	$\to Ar^1—NH_2 + Ar^2—NH_2$	
N-Oxides	$R^2—\overset{\displaystyle R^1}{\underset{\displaystyle R^3}{N}} \to O$	$\to R^2—\overset{\displaystyle R^1}{\underset{\displaystyle R^3}{N}}$	
Sulfoxides	$\overset{\displaystyle R^1}{\underset{\displaystyle R^2}{S}} \to O$	$\to \overset{\displaystyle R^1}{\underset{\displaystyle R^2}{S}}$	
Aldehydes, ketones	$\overset{\displaystyle R^1}{\underset{\displaystyle R^2}{C}}{=}O$	$\to \overset{\displaystyle R^1}{\underset{\displaystyle R^2}{CH}}—OH$	
Alkenes	$\overset{\displaystyle R^1}{\underset{\displaystyle R^2}{C}}{=}\overset{\displaystyle R^3}{\underset{\displaystyle R^4}{C}}$	$\to \overset{\displaystyle R^1}{\underset{\displaystyle R^2}{CH}}—\overset{\displaystyle R^3}{\underset{\displaystyle R^4}{CH}}$	

Sources: Refs. 1, 2, 42.

tion of carbonyl groups in aldehydes or ketones. Hydrogenation of alkenes produces saturated aliphatic hydrocarbons.

The hydrolytic reactions in the gut flora (Table 9) follow the same pathways as those mediated by the host enzymes, at least from a chemical point of view. For more information on reaction details the reader is referred to Table 5.

The importance of drug metabolism in the intestinal flora is illustrated by the actual examples compiled in Table 11. Although the list is far from complete, it shows that compounds belonging to many different chemical classes are liable to this type of bio-transformation, once they reach the colon. Following oral adminis-tration, some compounds such as the glucoside rutin or the azo

Table 11 Examples of Drugs that are Substrates for Enzymes of the Intestinal Microflora

Substrate	Reaction	Enzyme
Acetyl digoxin	Hydrolysis of ester bond	Esterase
Caffeic acid	Reduction of C=C-double bond	Hydrogenase
Chloramphenicol	Reduction of NO_2-group	Nitroreductase
Cyclamate	Cleavage of sulfamate bond	Sulfatase
Isonicotinuric acid	Hydrolysis of amide bond	Amidase
Metronidazole	Reduction of NO_2-group	Nitroreductase
Morphine glucuronide	Cleavage of glucuronide bond	Glucuronidase
Nicotine-N-oxide	Removal of O-atom in N → O-group	N-Oxide reductase
Nitrazepam	Reduction of NO_2-group	Nitroreductase
Oestradiol glucuronide	Cleavage of glucuronide bond	Glucuronidase
Oestrone	Reduction of keto group	Hydrogenase
Oestrone sulfate	Cleavage of sulfate ester bond	Sulfatase
Rutin	Cleavage of glucoside bond	Glucosidase
Sulfasalazine	Cleavage of N=N-double bond	Azoreductase
Sulfinpyrazone	Removal of O-atom in S → O-group	Sulfoxide reductase

Sources: Refs. 7, 9, 12, 14, 15, 45, 46

derivative sulfasalazine arrive in the colon because they are not absorbed in the upper intestine. Drug conjugates, such as morphine glucuronide or oestrone sulfate, are excreted in the bile and move unabsorbed to the large intestine (7, 9, 12, 14, 15, 45, 46).

The above examples reconfirm that microbial metabolism in the human gut proceeds primarily by reduction and hydrolysis. In several instances the products formed are biologically active.

C. Differences Between Metabolism in Gut Flora and Liver

The microbial flora in the colon has a metabolic potential equal to or greater than that of the liver (47). However, microbial and hepatic biotransformation differ from each other in several important aspects. All drugs systemically available will pass through the liver and will be exposed to the enzymes present there. Access to the colonic flora, on the other hand, is restricted to a few special cases, even after oral administration. Slow intestinal absorption, biliary excretion, and site-specific release from a delivery system are the three main reasons for a drug to reach those parts of the intestine where the microbial population becomes important.

In addition to these kinetic factors, the properties and activities of the enzymes have to be considered. Cytochrome P-450, which is the major drug-metabolizing enzyme system in the liver, catalyzes oxidative reactions and requires molecular oxygen. This type of phase I reaction is absent in the anaerobic environment of the gut flora in vivo.

Nitroreduction and some other reductive phase I processes are predominant in the flora and unimportant in the liver. The pertinent enzymes, although present in the hepatocytes, are largely inhibited by oxygen. Hydrolytic metabolism occurs at both sites, but seems to be more diversified and quantitatively more important in the flora than in the liver.

The liver is also rich in enzymes mediating the conjugation of drugs and drug metabolites with endogenous substituents. The micro-organisms in the large intestine do not contain conjugative enzymes. Since hydrolytic enzymes are abundant there, cleavage of conjugates is common.

From the above examples, it is obvious there are important qualitative differences between hepatic and microbial metabolism. If a drug gains specific access to the gut flora (e.g., by a suitable drug delivery system), it may undergo transformations that would not occur in the liver. In general, oxidation and conjugation are the preferred reactions in the liver while reduction and hydrolysis are the preferred ones in the gut lumen.

V. PRACTICAL IMPLICATIONS OF DRUG METABOLISM IN THE COLON

A. Pharmacologic and Toxicologic Consequences

The processes of phase I metabolism (i.e., oxidation, reduction, and hydrolysis) do not always lead to a loss of the biological activities of a xenobiotic, as already outlined above. The primary metabolites are often pharmacologically active, and may contribute to the therapeutic result of a medication. In some other instances, phase I metabolism gives rise to toxic or chemically reactive intermediates that may cause deleterious effects in the body. On the other hand, all conjugates formed by phase II metabolism are inactive, except in a few special cases. The liver is equally well equipped with enzymes for both phases of drug metabolism. This means that the end products of hepatic biotransformation are normally without biological effects.

Although conjugative inactivation does also occur in the colon, it is less important than in the liver. Conjugating enzymes are present in the enterocytes of the gut wall, but they are missing in the intestinal flora (Tables 7, 11). Microbial metabolism in the intestinal lumen is therefore, restricted to phase I reactions. They differ qualitatively from those occurring in the liver and may yield all kinds of potentially active products. Additional phase I reactions may take place in the gut wall. If one considers the potential of colonic drug metabolism as a whole, a variety of pharmacokinetic, pharmacologic, and toxicologic implications must be expected.

Illustrative examples of colonic metabolism with known or supposed toxicologic consequences are listed in Table 12. Except for drugs, the list also includes some xenobiotics that may be ingested as contaminants or ingredients of foodstuffs. Adverse local effects in the colon wall may be produced by substances that undergo oxidative metabolic activation in the enterocytes. For example, epoxides formed from aflatoxin B_1 or from polycyclic aromatic hydro-

Table 12 Examples of Colonic Metabolism with Toxicologic Implications

Substrate	Reaction	Site	Consequences
Aflatoxin B$_1$	C-Oxidation	Gut wall	Formation of reactive epoxide Binding to colonic DNA
Amygdalin	Glucoside cleavage	Gut flora	Liberation of cyanide Systemic cyanide toxicity
Azo food colors	Azoreduction	Gut flora	Formation of aromatic amines Systemic genotoxicity
Benzo(a)pyrene	C-Oxidation	Gut wall	Formation of reactive epoxide Binding to colonic DNA
Chloramphenicol	Nitroreduction	Gut flora	Formation of toxic metabolites Aplastic anemia
Cyclamate	Sulfamate cleavage	Gut flora	Formation of toxic metabolite Testicular atrophy
Metronidazole	Nitroreduction	Gut flora	Formation of toxic metabolites Systemic genotoxicity

Sources: Refs. 7, 9 14, 15, 45, 49, 50

carbons, such as benzo(a)pyrene, bind to colonic DNA and may be involved in carcinogenesis. Microbial metabolism of poorly absorbable xenobiotics often leads to local formation or liberation of lipophilic products readily absorbed through the colon wall. Such products may exert systemic toxicity. This mechanism explains, for instance, why metronidazole may cause aplastic anemia after oral but not after parenteral administration (7, 9, 14, 15, 45, 49, 50).

Pharmacokinetic and pharmacologic implications of colonic drug metabolism are also well documented (Table 13). Several drugs containing phenolic or alcoholic hydroxyl groups show reduced systemic availability following ingestion, because they undergo first-pass conjugation in the gut wall. Sulfation of isoproterenol and glucuronidation of morphine may serve as examples. However, inactive prodrugs such as pivamipicillin are activated by first-pass ester hydrolysis in the mucosa. Activation by oxidative metabolism is also known to occur as exemplified by ketodesogestrel, which is formed as an active metabolite of desogestrel in the mucosa (3, 5, 7, 12, 19, 45, 48, 51). Some drugs listed in Table 13 are activated by microbial instead of mucosal enzymes. Microbial cleavage of sulisatin and sulfasalazine results in laxative and antiinflammatory substances that exert their effects in the colon. Sulfinpyrazone is subject to microbial sulfoxide reduction. The result-

Table 13 Examples of Colonic Metabolism with Pharmacologic Implications

Substrate	Reaction	Site	Consequences
Desogestrel	C-Oxidation	Gut wall	Formation of active metabolite Enhanced systemic activity
Isoproterenol	O-Sulfation	Gut wall	Formation of conjugate Reduced systemic availability
Morphine	O-Glucuronidation	Gut wall	Formation of conjugate Reduced systemic availability
Pivampicillin	Ester hydrolysis	Gut wall	Cleavage of prodrug Enhanced systemic availability
Sulfasalazine	Azoreduction	Gut flora	Release of active moiety Local effect in colon
Sulfinpyrazone	Sulfoxide reduction	Gut flora	Formation of active metabolite Enhanced systemic activity
Sulisatin	Sulfate cleavage	Gut flora	Release of active moiety Local effect in colon

Sources: Refs. 3, 5, 7, 12, 19, 45, 48, 51

ing thioether that is only formed if part of an oral dose reaches the colon is absorbed and inhibits aggregation of blood platelets more strongly than the parent drug.

The example of sulfinpyrazone clearly shows that exposure of a drug to the colonic flora may lead to metabolites of unexpected properties. Biotransformation in the gut lumen or wall will, therefore, often have toxicologic and pharmacologic consequences different from those of hepatic metabolism. So far, however, colonic metabolism and its implications have mostly been evaluated in a retrospective and haphazard manner. The unique features of colonic metabolism have only rarely been considered in a prospective and systematic way, for example, in the design of drug molecules or in the development of pharmaceutical formulations.

B. Opportunities for Drug Design

Targeting of active substances to specific anatomical sites is a major aim of the design of modern drugs and delivery systems. Specific biotransformation in the large intestine offers two different possibilities of drug targeting: (1) a therapeutic effect can be selectively directed to and exerted at the colonic site and (2) a systemically active drug that would be unstable in the upper intestinal lumen can be released into the colon and absorbed from there.

The first of the two concepts has been translated into practice in a few cases, by making use of microbial metabolism. Inflammatory disease of the large bowel can be treated with oral doses of sulfasalazine. This compound is not absorbed in the upper gut and releases 5-aminosalicylic acid as the active moiety upon cleavage in the colon (Table 13). Glycosides of steroid drugs have been tested for the same purpose. Other examples of this concept are conjugated laxatives, which are inactive per se and split by the gut flora (10, 48, 52).

Attempts have been made to develop a suitable delivery system that would allow the release of any drug selectively into the colon. The crucial part of this experimental delivery system is an impermeable polymeric coat containing cross-linked azoaromatic groups. When the azopolymer-coated system reaches the large intestine, the azo groups are cleaved by microbial azoreductase, the polymer is broken down, and the drug is liberated (10, 48).

Peptides, such as insulin and vasopressin, are the primary candidates for azopolymer-coated systems. In the stomach and small intestine such peptides would be degraded by the digestive host enzymes in these parts of the alimentary canal. The colon is considered a favorable absorption site for peptides, because of the absence of digestive enzymes (section II). The "colonic coated" systems appear very attractive, but they are still in a conceptional phase of development, and more effort is needed before they will become available for use in patients.

The examples presented above prove that drug targeting to the colon is feasible in principle, if advantage is taken of the metabolic potential of the gut flora. However, this approach has not been fully exploited so far.

VI. CONCLUSIONS

Most of the conventional therapeutic drugs are rapidly absorbed after oral administration and removed from the gut lumen during their passage through the small intestine. However, sizable fractions of a dose may reach the large intestine, if a substance is poorly absorbed or released directly into the colon from a suitable drug delivery system. Such substances may undergo specific biotransformations in the large intestine.

Drug metabolism in the colon may be brought about by host enzymes in the enterocytes of the intestinal mucosa and by microbial enzymes in the gut flora. The colonic mucosa resembles the small intestinal mucosa regarding the spectrum of metabolizing enzymes. However, the total metabolic capacity of the colonic wall is inferior, since the mucosal mass in the lower part of the intestine is several times smaller than in the upper part. On the other hand, the gut flora, with its high metabolic capacity, is primarily found in the large intestine.

Metabolic reactions in the wall of the colon are oxidative, hydrolytic, and conjugative. They are catalyzed by cytochrome P-450, by esterases, amidases, and by various transferases. Reductive drug metabolism seems not to be important at this site. Enzyme activities in the colonic mucosa are inferior to those in the liver. The enzymes in the colon wall are readily saturated by high local drug concentrations.

The environment of the intestinal flora is strictly anaerobic. Therefore, no oxidative drug metabolism can take place in the lumen of the large intestine. Formation of conjugates is also unimportant since the intestinal flora contains no transferases. Reductive and hydrolytic processes occur instead, mainly mediated by nitro, azo, and carbonyl reductases, and by glucuronidases and glucosidases. Drug metabolism in the intestinal flora is generally more important than in the colon wall.

Metabolism in the large intestine may have toxicologic and pharmacologic consequences. Reactive intermediates formed by oxidation of xenobiotics in the mucosa may be involved in the development of colonic cancer. Microbial nitro and azo reduction may yield metabolites that exert systemic toxicity. Systemic availability of active substances may be decreased by first-pass glucuronidation or sulfation in the colonic mucosa. As an alternative, availability may be provided by cleavage of prodrugs in the gut wall or flora.

The unique enzymatic features of the intestinal flora lend themselves for drug targeting to the colon. Active moieties intended for local treatment can be liberated from poorly absorbable parent molecules by microbial processes such as reductive cleavage. Release of drugs from a delivery system disintegrating only by the action of a microbial reductase is also conceivable. The latter approach could be used for substances unstable in the upper part of the intestine, but are absorbed in the lower part.

REFERENCES

1. R. N. Armstrong, *CRC Crit. Rev. Biochem.* 22: 39 (1987).
2. B. Testa, P. Jenner, *Drug Metabolism: Chemical and Biochemical Aspects*. Marcel Dekker, New York, (1976).
3. D. J. Back, S. M. Rogers, *Aliment. Pharmacol. Ther* 1: 339 (1987).
4. J. Caldwell and M. Varwell Marsh, in *Butterworths International Medical Reviews, Clinical Pharmacology and Therapeutics 1. Presystemic Drug Elimination* (C. F. George, D. G. Shand, and A. G. Renwick, eds.) Butterworth Scientific, London, (1982) p. 29.
5. C. F. George, *Clin. Pharmacokinet.* 6: 259 (1981).
6. O. Hänninen, P. Lindström-Seppä, K. Pelkonen, *Arch. Toxicol.* 60: 34 (1987).
7. K. F. Ilett, L. B. G. Tee, P. T. Reeves, R. F. Minchin, *Pharmacol. Ther.* 46:67 (1990).
8. M. Laitinen, J. B. Watkins, in *Gastrointestinal Toxicology*, (K. Rozman and O. Hänninen, eds.) Elsevier Science, Amsterdam (1986), p 169.

9. A. G. Renwick, in *Butterworths International Medical Reviews, Clinical Pharmacology and Therapeutics 1. Presystemic Drug Elimination* (C. F. George, D. G. Shand, and A. G. Renwick, eds.) Butterworth Scientific, London, (1982), p 3.
10. W. A. Ritschel, *Methods Find. Exp. Clin. Pharmacol* 13: 313 (1991).
11. M. J. Iatropoulos, in *Gastrointestinal Toxicology*, (K. Rozman and O. Hänninen, eds.) Elsevier Science, Amsterdam (1986), p 246.
12. H. G. Boxenbaum, I. Bekersky, M. L. Jack, S. A. Kaplan, *Drug Metab. Rev.* 9: 259 (1979).
13. B. R. Goldin, *Ann. Med.* 22: 43 (1990).
14. I. R. Rowland, *Drug. Metab. Rev.* 19: 243 (1988).
15. I. R. Rowland, A. Wise, *CRC Crit. Rev. Toxicol.* 16: 31, (1985).
16. D. C. Taylor, *Pharm. Int.* 179 (1986).
17. H. P. Hoensch, *Banbury Rep.* 11: 89 (1982).
18. H. P. Hoensch, R. Hutt, F. Hartmann, *Environ. Health Perspect.* 33: 71 (1979).
19. H. P. Hoensch, M. Schwenk, in *Intestinal Toxicology* (C. M. Schiller, ed.) Raven Press, New York, (1984) p 169.
20. O. Hänninen, *Arch. Toxicol. Suppl.* 8: 83 (1985).
21. W. H. M. Peters, P. G. Kremers, *Biochem. Pharmacol.* 38: 1535 (1989).
22. H. W. Strobel, S. N. Newaz, W.-F. Fang, P. P. Lau, R. J. Oshinsky, D. J. Stralka, F. F. Salley, in *Extrahepatic Drug Metabolism and Chemical Carcinogensis* (J. Ryström, J. Montelius, and M. Bengtsson, eds.) Elsevier Science, Amsterdam (1983) p 57.
23. A. S. Koster, in *Progress in Pharmacology and Clinical Pharmacology*, vol. 7/2. Gustav Fischer Verlag, Stuttgart (A. S. Koster, E. Richter, F. Lauterbach, and F. Hartmann, eds.) New York (1989) p 181.
24. G. M. Pacifici, M. Franchi, C. Bencini, F. Repetti, N. Di Lascio, G. B. Muraro, *Xenobiotica* 18: 849 (1988).
25. G. M. Pacifici, A. Temellini, L. Guiliani, A. Rane, H. Thomas, F. Oesch, *Arch. Toxicol.* 62: 254 (1988).
26. K. F. Ilett and D. S. Davies, in *Butterworths International Medical Reviews. Clinical Pharmacology and Therapeutics 1. Presystemic Drug Elimination* (C. F. George, D. G. Shand, and A. G. Renwick, eds.) Butterworth Scientific, London (1982) p 43.
27. G. M. Pacifici, C. Bencini, A. Rane, *Xenobiotica* 16: 123 (1986).
28. G. M. Pacifici, L. Giuliani, R. Calcaprina, *Pharmacology* 33: 103 (1986).
29. G. M. Pacifici, M. Franchi, C. Colizzi, L. Giuliani, A. Rane, *Pharmacology* 36: 411 (1988).
30. G. M. Pacifici, M. Franchi, L. Giuliani, *Pharmacology* 38: 146 (1989).
31. W. H. M. Peters, P. L. M. Jansen, *Biochem. Pharmacol.* 37: 564 (1988).
32. M. Cappiello, M. Franchi, L. Giuliani, G. M. Pacifici, *Eur. J. Clin. Pharmacol* 37: 317 (1989).

33. M. Cappiello, L. Giuliani, G. M. Pacifici, *Pharmacology* 40: 69 (1990).
34. M. Schwenk, in *Progress in Pharmacology and Clinical Pharmacology*, vol. 7/2. Gustav Fischer Verlag, Stuttgart (A. S. Koster, E. Richter, F. Lauterbach, and F. Hartmann, eds.) New York (1989) p 155.
35. G. M. Pacifici, M. Franchi, P. G. Gervasi, V. Longo, P. di Simplicio, A. Temellini, L. Giuliani, *Pharmacology* 38: 137 (1989).
36. A. J. Baars, H. Mukhtar, C. E. M. Zoetemelk, M. Jansen, D. D. Breimer, *Comp. Biochem. Physiol.* 70C: 285 (1981).
37. G. Batist, K. Mekhail-Ishak, N. Hudson, J. M. DeMuys, *Biochem. Pharmacol.* 37: 4241 (1988).
38. P. C. Hayes, D. J. Harrison, I. A. D. Bouchier, L. I. McLellan, J. D. Hayes, *Gut* 30: 854 (1989).
39. W. H. M. Peters, F. M. Nagengast, T. Wobbes, *Carcinogenesis* 10: 2371 (1989).
40. W. H. M. Peters, H. M. J. Roelofs, F. M. Nagengast, J. H. M. van Tongeren, *Biochem. J.* 257: 471 (1989).
41. C.-P. Siegers, in *Progress in Pharmacology and Clinical Pharmacology*, vol. 7/2. Gustav Fischer Verlag, Stuttgart (A. S. Koster, E. Richter, F. Lauterbach, and F. Hartmann, eds.) New York (1989) p 171.
42. J. W. Faigle, in *Biopharmazie* (J. Meier, H. Rettig, and H. Hess, eds.) George Thieme Verlag, Stuttgart/New York (1981) p 147.
43. A. S. Koster, in *Advances in Glucuronide Conjugation*, Falk Symposium 40 (S. Matern, K. W. Bock, and W. Gerock, eds.) MTP Press, Lancaster (1985) p 177.
44. W. H. M. Peters, F. M., Nagengast, J. H. M. van Tongeren, *Gastroenterology* 96: 783 (1989).
45. P. Goldman, in *Microbial Transformations of Bioactive Compounds*, *vol. 2* (J. P. Rosazza, ed.) CRC, Boca Raton (1982) p 43.
46. H. P. A. Illing, *Xenobiotica* 11: 815 (1981).
47. R. R. Scheline, *Pharmacol. Rev.* 25: 451 (1973).
48. A. Rubinstein, *Biopharm. Drug Dispos.* 11: 465 (1990).
49. H. Autrup, *Drug Metab. Rev.* 13: 603 (1982).
50. H. Autrup, R. C. Grafstrom, in *Developments in Biochemistry, vol. 23. Cytochrome P-450, Biochemistry, Biophysics and Environmental Implications* (E. Hietanen, M. Laitinen, and O. Hänninen, eds.) Elsevier Biomedical, Amsterdam (1982) p 643.
51. S. Madden, D. J. Back, C. A. Martin, M. L'E. Orme, *Br. J. Clin. Pharmacol.* 27: 295 (1989).
52. T. N. Tozer, J. Rigod, A. D. McLeod, R. Gungon, M. K. Hoag, D. R. Friend, *Pharm. Res.* 8: 445 (1991).
53. H. W. Davenport, in *Physiology of the Digestive Tract* (Year Book Medical Publishers, Chicago (1966) p 170.
54. *The New Encyclopaedia Britannica. Macropaedia, vol. 5*, Encyclopaedia Britannica, Chicago/London (1982) p 793.

3

In Vitro Studies with Colonic Tissue, Cellular, and Subcellular Preparations

Pierre Dechelotte*and Michael F. Schwenk†

Medizinische Hochshule, Hannover, Germany

I. INTRODUCTION

As a correlate to the development of sustained-release drug formulations, an increasing need for better knowledge of the fate of drugs in the gastrointestinal tract has arisen in the past decade. This general trend is most evident in the case of the colon. Its remote aboral location has prompted the use of clinical methods control ling the site-specific delivery of drugs (intubation, endoscopy scintigraphy, capsules, and other methods; see Chapters 5 and 6). The contraluminal side of the colonic mucosa in humans is inaccessible. However, it is necessary to characterize its role in drug transport and metabolism apart from that of the associated flora. Therefore, in vitro models for exploration are of crucial importance. Furthermore, in vitro methods allow to identify both the drug-metabolizing cell types and the responsible enzymes. This is of great importance for the assessment of the role of colonic biotransformation in colonic carcinogenesis.

---
* *Present affiliation*: Policlinique–GBPDN, University Hospital, Rouen, France

† *Present affiliation*: Department of Environmental Toxicology, Federal Health Agency, Stuttgart, Germany

Table 1   Advantages and Limits of In Vitro Models

Models	Comments
Colonic slices	Simple method for screening of uptake and metabolism. Poor reproducibility of tissue slicing and of O_2 supply.
Biopsies	Histochemical and biochemical identification of enzymes and their regional distribution. Limitations due to small size and unknown viability.
Everted sacs	Simple assessment of metabolism and transport of both parent drug and formed metabolites. Unphysiological serosal volume and oxygenation.
Perfused segments	Variable rates of perfusion for kinetic studies. Complex experimental procedure.
Diffusion chambers	Assessment of transport of drug and its formed metabolite(s) through a single sheet of isolated mucosa. Electrical or biochemical control of integrity required.
Organ culture	Long-term maintenance of cellular integrity and epithelial architecture, allowing prolonged study of drug metabolism and cellular effects of metabolites. Complex procedure, not suitable for transport studies.
Isolated colonocytes	Smallest intact entity allowing study of uptake and metabolism with good control of extracellular conditions. Loss of mucosal–serosal polarity.
Cell lines	Suitable for transport study if monolayer is formed. Derived from carcinomatous cell lines with varying differentiation. Drug metabolizing enzyme activities still poorly understood.
Homogenates	Allow screening of some biotransformation reactions. Nonselective mixing of material from different cell types and subcellular fractions. Possible alterations of enzymatic activities related to procedure.
Subcellular preparations	Standard procedure to assess mixed function oxidase and other enzyme activities. Provide information on enzyme kinetics and subcellular localization. Membrane vesicles suitable for transport studies, but loss of intracellular regulatory processes.

This chapter focuses on in vitro models available for studies with material from either animal or human origin, in order to explore drug transport through and metabolism by the colonic mucosa. A summary of these methods with their respective advantages and limits is given in Table 1. In vitro studies with small intestinal mucosa are mentioned only for comparison with the colon. The interested reader is referred to some recent reviews (1–5).

II. GENERAL PRINCIPLES OF IN VITRO STUDIES

In all experimental models some principles should be followed.

1. The colon must be cleaned to remove feces and bacteria by gentle rinsing the lumen with an adequate amount of buffer. In some animal species (e.g., the guinea pig) the buffer must be prewarmed above 30° C to avoid colonic constrictions.
2. The colonic epithelium is very sensitive. Any traumatic damage will release proteolytic enzymes from lysosomes, which will promote autolysis, unless protease inhibitors are added. It is, therefore, most important to work with greatest caution and to remove debris by occasional intermittent washings. All steps must be performed under oxygenation with 100% oxygen or with carbogen (95% oxygen, 5% CO_2). Even under such conditions, the deepest layers of the tissue specimen may not be sufficiently supplied with oxygen.
3. The preparation is incubated with suitable medium at 37° C under oxygenation. This may be simple Krebs–Henseleit buffer or a more elaborate tissue culture medium containing a mixture of nutrients. Addition of antibiotics, antifungals, albumin (0.1–0.5%), or fetal serum is recommended for prolonged incubation and cultures.

III. STUDIES WITH COLONIC TISSUE
A. Some Physiological Considerations

The colonic epithelium forms a barrier between lumen and blood stream (Fig. 1). Therefore, drug absorption may occur either by diffusion across the tight junctions (paracellular route) or by permeation through the epithelial cell (transcellular route). The sec-

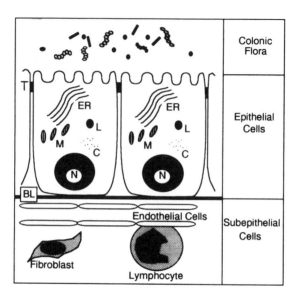

Figure 1 Sites of colonic biotransformation. Colonocytes are connected with neighboring cells by tight junctions (T), sealing intercellular space from lumen. Cells are anchored to the basal lamina (BL), which is permeable to drugs. Subepithelial cells may participate in drug metabolism, but colon epithelial cells seem to be most important. Biotransformation in the epithelial cell is mainly oxidative and conjugative, while that in the colonic flora is mainly reductive and deconjugating. Major intracellular drug-metabolizing sites are the endoplasmic reticulum (ER), cytosol (C), mitochondria (M), lysosomes (L), and nucleus (N).

ond route involves cellular uptake, possible intracellular biotransformation, and release across the basolateral membrane into the subepithelial space. From there, the drug enters the bloodstream. In preparations of colonic tissue, the continuity of the epithelial lining is preserved. Nonepithelial cell types are present (Fig. 1). There are bacteria sticking to the surface, subepithelial cells (e.g. endothelial cells, fibroblasts, lymphocytes), and cells of the muscularis and serosa. Nonepithelial cell types may contribute to biotransformation reactions, but will sometimes "dilute" the enzyme activity of the epithelial cells. Moreover, the subepithelial layers constitute an artificial transport barrier. In this chapter we define

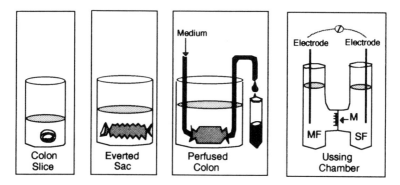

Figure 2 Experimental models with colonic tissue. Schematic drawing of four in vitro methods. In each, the tissue is bathed in suitable medium. Devices for oxygenation and temperature control are not shown. M, mucosal sheet; MF, mucosal fluid; SF, serosal fluid.

transport as transepithelial movement, as transmucosal movement if the mucosa is not stripped, uptake as movement into the epithelial cell, and release as movement out of the epithelial cell.

The following section reviews studies using tissue of the colonic wall in vitro. Some important models are illustrated in Figure 2: colonic slices, colonic sacs, and perfused colonic epithelial sheets in diffusion chambers (Ussing chambers). Studies with colonic tissue are listed in Table 2.

B. Colonic Slices and Biopsies

Colonic slices are gained by cutting rings without damage from the freshly excised and carefully rinsed colon of experimental animals. These slices can be immediately used for incubations. They represent the simplest in vitro method suitable to attain information about drug metabolism and drug transport.

Metabolites are measured in the incubation medium, while drug uptake is assessed by measuring drug levels in the tissue. The simplifying assumption is made that during the initial phase of uptake, the drug predominantly accumulates within epithelial cells. Colonic slices have been used in early studies on the distribution of conjugative or oxidative enzymatic systems along the digestive

Table 2 Studies with Colonic Tissue Preparations

Models	Species	Compounds	Transport	Metabolism	References
Colonic slices	Rat	o-Aminophenol		G	8
Biopsies	Human			Cytochrome P450	11
	Dog			UDP-gDH	9
Everted sacs	Chicken	Uracil	Active T		17
	Rat	Thyroxine, tri-iodothyronin analogs	T	G	18
	Rat	Insulin	T	Degradation	19
	Rat	Laxative diphenols	T and U	G	21
	Rat	Bisacodyl	T	Hydrolysis	22
	Rat	Danthron, rhein	T	G and S	23
	Rat	Inulin + promoters	T		20
Ussing	Rat	Magnesium	T		26
chambers	Rabbit	5-fluorouracil	T	Conjugation	25
	Chicken	Uracil	Active T		17
	Guinea pig	Pirenzepine,	Secretion		33
		Telenzepine	Secretion		33
	Guinea pig	Naphthol,	T	G and S	35, 36
		phenolphthalein	T	G and S	35
	Guinea pig	Cymarol,	Secretion		34
		strophantidol	Secretion		34
	Guinea pig	Estradiol	T	Conjugation	37
	Guinea pig	p-Amino benzoic acid	T	Metabolism	38
Organ	Human	Benzo(a)-pyrene	Binding	Metabolism	43–46
cultures	Human	DMN, DMH	Binding	Metabolism	44
	Human	1-naphthol		G and S	46, 47

T, transport; U, mucosal uptake; G, glucuronide; S, sulfate conjugation; DMN, dimethylnitrosamine; DMH, dimethylhydrazine; UDP-gDH, uridine diphosphate glucose dehydrogenase.

tract (6–9) and in different parts of the large intestine (for review, see 10). The lowest conjugation rate measuring aminophenol glucuronidation was observed in slices from the cecum (8). The conjugation rate was slightly higher in the colon, but still only 30% of the duodenal rate. On the contrary, colonic slices exhibited the highest β-glucuronidase activity and cecal slices showed UDP–glucose dehydrogenase activities as high as the slices from the duodenum.

Fixed slices have been used to localize with histochemical techniques the enzymes involved in glucuronidation in colonic epithelium (9). Immunohistochemical techniques will be of increasing importance in the near future. They allow the localization of drug-metabolizing isozymes, such as cytochrome P450 isozymes, even at very low levels with high specificity (11). Although these methods are extremely valuable for the study of enzyme levels and localizations, they do not give information about the enzyme activity towards a certain drug in living tissue.

Fresh surgical or endoscopic colonic biopsies can be handled in the same way as colonic slices. However, there are several unknowns, such as the extent of trauma and the exact composition (mucosa vs muscularis). Thus, biopsies are usually not used directly, but to isolate cells or subcellular fractions (see Sections IV and V).

C. Colonic Segments

Small intestinal segments have been used in various ways as everted or noneverted sacs (12), luminally perfused segments (13, 14), luminally and vascularly perfused segments (15, 16). In contrast, colonic segments have been almost exclusively used as everted sacs (see Table 2 for references).

Everted sacs are adequate to assess the transport of luminally or serosally administered drugs (17–20) and the release of their metabolites on both sides (21–23) because the polarity is maintained. An active transport can be characterized, for example, for uracil (17) and thyroxine analogs (18), if the equilibrium drug concentration on the serosal side exceeds that on the mucosal side. Furthermore, the effects of permeants on transcellular and paracellular pathways can be assessed (20). Serial studies with diphenol (21, 22) and anthraquinone laxatives (23) have established that these compounds are quantitatively metabolized during their transport across rat colonic sacs. There is subsequent accumulation of glucuronides on the serosal side and minute secretion of sulfate back into the lumen for anthraquinones (23). Likewise, less than 1% of administered radiolabeled insulin crossed unchanged everted and noneverted sacs of rat distal colonic epithelium. The major fraction was degraded by either apical or submucosal layers (19).

Although they yield quantitative and reproducible data, everted sacs may be criticized. The eversion procedure results in unphysiological disproportions between the new mucosal and serosal compartment, the former being too wide and the latter inadequately confined and oxygenated. Moreover, the accumulation of transported drugs or metabolites on the serosal side, a situation remote from the high physiological splanchnic outflow, may rapidly exert an inhibitory effect on the transepithelial movements. The contribution of the mucosa by itself to the observed metabolism cannot be differentiated from that of other tissue layers. It may even be partially blunted because of poor diffusion of the formed metabolites across subepithelial tissue.

D. Isolated Mucosa in Diffusion Devices

Several of the limits of everted sacs may be avoided by working with sheets of isolated colonic mucosa mounted in diffusion devices (Fig. 1), such as Ussing chambers (24–26). Ussing chambers were initially developed to study transepithelial ionic fluxes and the associated electrophysiological changes in various parts of the gastrointestinal tract including the human colon (27). The technique has been later adapted for studies of drug transport and metabolism in the small intestine (28–30) and stomach (31).

Immediately after removal or surgical resection, the mucosal and serosal layers are stripped off. The isolated mucosa is mounted between two reservoirs that provide bathing of "mucosal" and "serosal" sides by equal volumes of medium with adequate temperature, oxygenation and stirring (Fig. 2). Unidirectional transfers of drugs (mucosa to serosa and serosa to mucosa) can be assessed, as well as the release of metabolites on both sides. The viability of the preparation is monitored either by continuous recording of the transepithelial potential difference (10–20 mV for human colon) (27) or by checking the absence of permeation of mucosally added labeled polyethylene glycol and/or inulin (29). Using Ussing chambers has the distinct advantage of reliable data collection on transport and metabolism, even with small specimens of human colon.

Both mucosa to serosa and serosa to mucosa fluxes of fluorouracil through rabbit proximal and distal stripped colon (25) have been shown to increase linearly with dose. In contrast, ileal mucosa to serosa flux showed a saturable component in addition. Fluorouracil was metabolized during transport to more polar compounds transported in the same direction as the parent drug. An active sodium-dependent mucosa to serosa transport of uracil, which was inhibited by hypoxanthine, has also been characterized in chicken colon and jejunum (17).

Lauterbach et al. use a modified diffusion device (29). They have shown in several studies that some compounds are actively secreted (32–34). This was most strikingly illustrated with the non-quaternary amines pirenzepine and telenzepine on guinea pig stripped colon. The serosa to mucosa flux of telenzepine was 3–18 times higher than the corresponding mucosa to serosa flux (33). This secretion was inhibited by phenoxybenzamine. The symmetrical set up of diffusion chambers is also appropriate to address the compartmentalization of conjugation reactions in intestinal and colonic mucosa (35–38; for review, see 39). Naphthol and phenolphthalein, for example (35), are glucuronidated and sulfated after both luminal or serosal administration. The site of release of the conjugates depends in the jejunum, but not in the colon, on the side of drug administration. However, compartmentalization has been observed for the N-acetylation of p-aminobenzoic acid in the colon (38).

E. Human Colonic Organ Culture

Assessing the effects of carcinogens and their metabolites on the colonic mucosa is of outstanding interest in the field of carcinogenesis (40–42). Colonic sacs or isolated colonic mucosa in Ussing chambers do not reliably survive beyond 2 or 3 hr. Therefore, they are not suitable for such studies. A method of explant culture of colon has been developed (43). It uses tissue obtained from autopsy or surgical specimens. After the muscular layers are dissected, the mucosal layers were cultivated in plastic dishes. After several improvements of incubation conditions, explants remained viable for up to 20 days, with good histologic maintenance of absorptive and goblet cells in the first 10 days. This model is useful for the study of slowly metabolized xenobiotics.

The ability of the cultured human mucosa to activate the procarcinogens benzo(a)pyrene, dimethylnitrosamine, or dimethylhydrazine to a wide variety of metabolites was observed in both normal and tumorous tissues (43–46). The binding of benzo(a)pyrene to DNA was highest in the descending colon, and could be enhanced by addition of taurodesoxycholic and lithocholic acid. It was also slightly higher in nontumorous tissues from patients with cancer than in tissues from patients without cancer (45). Other shifts of drug-metabolizing activities in tumorous colon have been observed. For example, 1-naphthol is preferentially sulfated by normal colon and preferentially glucuronidated by tumorous colon (46, 47).

IV. STUDIES WITH CELLULAR PREPARATIONS

A. Isolated Colonic Cells and Crypts

1. Specific Advantages of Isolated Colonic Cells

Isolated epithelial cells (colonocytes) have several advantages compared with tissue preparations. First, they do not contain cell types of other layers. Second, there is no problem with insufficient oxygen supply. Third, many incubations can be done in parallel and cell suspensions can be pipetted as homogeneous material. Isolated cells have also an advantage over subcellular preparations. They produce the entire pattern of drug metabolites. Thus, cellular preparations are invaluable tools to study metabolic pathways. They allow for a comparison between organs under reliable conditions. The epithelial lining is disrupted. Therefore, the study of drug transport with isolated cells is limited to its initial and final steps, namely the uptake and release of the drug. Whether the uptake takes place at the apical or basolateral site cannot be determined.

Isolated epithelial cells from both small and large intestine are different from cells of other tissues, such as stomach and liver. Their viability decreases very rapidly. This restricts their use to short-term incubations (1–2 hr).

2. Preparation of Isolated Colonic Cells and Crypts

Numerous methods have been proposed to prepare isolated colonic cells or crypts from animal or human tissue pieces (Table 3). For enterocytes (48), these methods differ by the use of several combinations of mechanical, chelation, and enzymatic procedures to separate the epithelial sheet from underlying tissue and to dissociate the monolayer epithelium. A mixture of single cells and crypt fragments is usually obtained. None of the diverse isolation procedures

Table 3   Colonic Cell and Crypt Isolation Procedures

Mechanical	Chemical	Enzymes	Species	References
Stirring, centrifugation			Human	63
Trituration	EDTA, citrate, DTT		Guinea pig	65
Disaggregation	EDTA, DTT		Rat	64
Scraping, pipetting	mannitol		Rat	66, 67
Rotation		Pronase	Rat	68
Vibration	EDTA		Rat	52
Disaggregation	EDTA, DTT	Hyaluronidase	Human	64
Disruption	Ca^{2+}-free medium	Trypsin	Human	69
Agitation	EDTA		Human	59
Subepithelial injection	EDTA, citrate, DTT		Human	61
Trituration		Dispase, collagenase	Human	70

EDTA, ethylene-diaminetetraacetic acid; DTT, 1,4-dithiothreitol.
The last two procedures (61, 70) relate to isolation of crypts.

has emerged as a gold standard. The success in preparing enough cells with good initial viability depends more on experience and routine than on the method employed.

A critical aspect of any cell isolation procedure is to assess routinely the viability of the cells before their use. In some cases, such as prolonged incubation or exposure to cytotoxic agents, repeated assessment of viability is mandatory. The most widely used tests of viability include oxygen consumption, lactate dehydrogenase (LDH) release, incorporation of radiolabeled uridin, and phenylalanine and microscopic examination combined with trypan blue exclusion test. In addition, electron microscopic examination should be done during the development of a new procedure, to check the integrity of membranes and intracellular organelles (Fig. 3) and to identify the various cell types (absorptive cells, goblet cells). Cell yield of the preparation is assessed by protein content or cell count. Further information concerning the evaluation of epithelial cell procedures may be found in references 48 and 49.

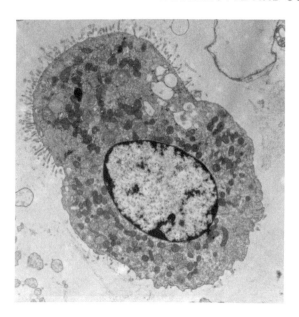

Figure 3   Electron micrograph of an isolated colonic cell. Cells were isolated with a calcium chelation method from adult guinea pig colon. Colonocytes show short microvilli and well preserved ultrastructure (from ref. 65).

3. Assessment of Biotransformation Reactions

Besides their repeated use for study of cellular energy metabolism (50–52), isolated colonic cells have been used successfully to study phase I (1, 53, 54) and phase II (55–60) biotransformation reactions (Table 4). Isolated human colon crypts give results similar to those of cellular preparations (61).

The cytochrome P450 content as well as aminopyrine demethylase and nitrosodimethylamine demethylase activities have been found to be much lower in colonocytes than in hepatocytes of the guinea pig (1, 65). In the same species, however, the colon carcinogen dimethylhydrazine (DMH) is demethylated in colonocytes at a similar rate as in hepatocytes, and even fourfold faster than in jejunal enterocytes. Ethoxycoumarin de-ethylase activity of colonocytes ranged between that of hepatocytes and enterocytes (1).

Table 4 Studies with Isolated Cells, Crypts, and Cell Lines

Models	Species	Compounds	Uptake/metabolism	References
Isolated	Rat	o-Aminophenol	G	8
colonocytes	Rat	1-Naphthol, morphine	G	55
	Guinea pig	Aminopyrine	Demethylation	65
	Guinea pig	Ethoxycoumarin	De-ethylation	1
	Guinea pig	Nitrosodimethylamine	Demethylation	1
	Guinea pig	1-Naphthol	G and S	56
	Guinea pig	Tyramine	Deamination	53
	Guinea pig	S- and R-naproxen	G and S	57
	Guinea pig	DMH	Demethylation	1
	Guinea pig	DMH	U, metabolism	62
	Hamster	DMH	U, metabolism	54
	Human	5-ASA	Acetylation	58, 59
	Human	5-ASA, N-acetyl ASA	U, acetylation	60
Isolated	Human	1-Naphthol	G and S	61
crypts	Human	Estrone, estradiol	G and S	61
	Human	DMH	Demethylation	61
Human	COLO 205	1-Naphthol	G and S	47
tumor cell	LOVO	1-Naphthol	G and S	47
lines	Caco-2	Cobalamin	U	73

U, uptake; G, glucuronide; S, sulfate conjugation; DMH, dimethylhydrazine; ASA, aminosalicylate; UDP-g, uridine diphosphate glucose; UDP-gDH, UDP-glucose-dehydrogenase; UDP-GA, uridine diphosphate glucuronic acid; UDP-GT, UDP-glucuronosyl transferase; PPHG, NG, MG, phenolphthalein, naphthol, and morphine glucuronides; β-G, β-glucuronidase; GSH, glutathione; GSH-T, GSH-S-aryltransferase; GSH-P, GSH-peroxidase.

A DMH demethylase activity was also quantifiable in human colonocytes (61). These data suggest the presence of specific isoenzymes in the colon activating DMH to the ultimate carcinogens. Monoamine oxidase is an other phase I enzyme, deaminating, among others, tyramine. Tyramine deamination activity in guinea pig colonocytes was lower than in hepatocytes or gastric parietal cells but considerably higher than in enterocytes (53).

Conjugating enzymes have varying activities in colonocytes, approaching in some cases those in hepatocytes. This has been demonstrated with substrates like 1-naphthol (55, 56, 61), morphine (55), estrogens (61), naproxen (57), and aminosalicylate (58–60). The velocity of 1-naphthol conjugation in guinea pig colonocytes is lower than in enterocytes, mainly due to reduction of glucuronidation, sulfatation being maintained (56). A similar pattern was also observed in human colonocytes (61). In the rat, glucuronidation activity for naphthol and morphine also decreases from the duodenum towards the ileum, but then increases along the cecum, colon, and rectum (55). Naproxen is poorly glucuronidated in colonocytes (3% of rate in hepatocytes). It is interesting that colonic cells preferentially glucuronidate the S-stereoisomer of the drug while liver cells glucuronidate the R-stereoisomer (57). Human colonocytes conjugate estrone and estradiol (61) and acetylate 5-aminosalicylate (58, 59).

4. Assessment of Drug Uptake

Although isolated colonic cells are suitable for uptake studies, only a few data have been gathered. Dimethylhydrazine was taken up by isolated colonocytes of the guinea pig. Intracellular concentrations exceeded the extracellular concentration by fivefold (62). Uptake of DMH was also documented in hamster colonic cells (54). In human colonocytes, the uptake of exogenous N-acetylaminosalicylate was much lower than that of the nonacetylated 5-aminosalicylate (60).

B. Cultured Colonic Cell Lines

1. Some Features of Cell Lines

Many investigators failed to establish primary cultures of colonic cells. It is not known whether this difficulty is due to loss of attachment receptors or to the short life span of cells. Despite some positive reports (70), primary cultures of colonic cells are therefore not routinely available for pharmacologic studies. As an alternative, some permanent cell lines originating from colonic adenocarcinoma tissues have been established and successfully subcultured (71). Such cell lines develop a various degree of differentiation, which can be partly influenced by culture conditions (72). Among others, Caco-2 and HT29 cell lines express some brush border enzyme activities typical of fetal human colon. T84 cells do not express microvillar hydrolases and are morphologically similar to adult colonic crypt cells. Caco-2 and T84 cell lines spontaneously form well-polarized monolayers with electrical parameters and ion transport activities analogous to those of adult colonic crypt cells.

In contrast, HT29 cells need special culture conditions to form a monolayer and can be subcloned to either multipotent or highly differentiated absorptive or goblet cells.

2. Studies of Drug Transport and Metabolism

Although cell lines have been repeatedly used by investigators to study differentiation and ion transport processes, data concerning their ability to transport and metabolize drugs and xenobiotics are still sparse. Caco-2 cell monolayers have been shown to bind and internalize both cobalamine alone and its complex with intrinsic factor (73). Siegers et al. (80, 81) noted that the Caco-2 cell line is also suitable to study biotransformation reactions. The total content of glutathione in Caco-2 cells was somewhat lower than in surgical biopsies from both normal and tumorous human colon. In contrast, the glutathione-S-aryltransferase and -peroxidase activities of this cell line exceeded by three to four times those of human biopsies and dimethylhydrazine demethylase activity of Caco-2 cells ranged between that of normal and tumorous human tissues. Three human colonic carcinoma cell lines (COLO 205, COLO 206, and LoVo) conjugated 1-naphthol almost exclusively to glucuronide and little if any to sulfate. Such a pattern is similar to that observed with explant culture and xenograft of human colonic tumors (47).

V. STUDIES WITH SUBCELLULAR PREPARATIONS

A. Advantages and Limits of Subcellular Preparations

The widespread use of subcellular fractions is related to some of their distinct advantages. Large quantities of material can be obtained in one processing and stored frozen for a relatively long time. Optimal levels of coenzymes or cosubstrates may be achieved. Due to the predominant localization of enzymes in certain organelles, activities can be enriched by cell fractionation. In fact, monoamine oxidase is localized in mitochondria, mixed function oxidase, and glucuronosyltransferase in microsomes, sulfotransferase and glutathionyltransferase in the cytosol, and β-gluc-

Table 5 Studies with Homogenates and Subcellular Fractions

Models	Species	Compounds	Metabolism	References
Homogenates	Rat	Benzo(a)pyrene	Hydroxylation	75
	Rat	Morphine	G	78
	Pig, rabbit	Dextran–naproxen	Dextranase	91
	Human	Morphine	G	78
	Human	5-ASA	Acetylation	59, 77
	Human	Benzo(a)pyrene	Oxidation	76
	Human	GSH	GSH-T, GSH-P	80
	Human	DMH	Demethylase	81
		Aldrin	Epoxidase	81
Subcellular fractions	Rat	Benzo(a)pyrene	Hydroxylation	79, 82, 87, 88, 92, 93
	Rat	Benzphetamine	Hydroxylation	82, 87, 92, 93
	Rat	Ethylmorphine	Dealkylation	82, 87, 92
	Rat	2-Aminoanthracene	Metabolic activation	87, 90
	Rat	DMH	Metabolic activation	89, 94
	Guinea pig	1-Naphthol, Morphine, 1-NG, 3-MG	UDP-gDH, UDP-GT, β-G	55
	Human	Phenol, estrone	Sulfokinase	83
	Human	1-Naphthol	G and S	47
	Human	Morphine	G	78
	Human	5-ASA	Acetylation	59
	Human	DMH	Oxidation	95
	Human	Benzo(a)pyrene	Hydroxylation	95
	Human cell-lines	Purine	Metabolic enzymes	84

Studies labeled "homogenates" were performed with rough homogenates or S9 fractions (see text for details). Studies labeled "subcellular fractions" used one or more nuclear, mitochondrial, microsomal, or cytosolic fraction.

β-G, β-glucuronidase; UDP-g, uridine diphosphate glucose; UDP-gDH, UDP-glucose-dehydrogenase; G, glucuronide; and S, sulfate conjugation; 5-ASA, 5-aminosalicylate; GSH, glutathione; GSH-T~, GSH-S-aryltransferase; GSH-P, GSH-peroxidase; DMH, dimethylhydrazine; 1-NG, 3-MG, 1-naphthol and morphine-3 glucuronides.

uronosyltransferase in lysosomes (Fig. 1). Thus, subcellular fractions are most suitable for enzymatic studies (Table 5). Moreover, membrane vesicles of both brush border and basolateral membranes are useful to study mechanisms of uptake and release. Yet the loss of many intracellular regulating processes due to cell disruption precludes any extensive extrapolation to the in vivo situation.

B. Preparation of Homogenates and Subcellular Fractions

Homogenates and subcellular fractions from colonic mucosa are prepared following the same stepwise procedure as for small intestine. This has been recently reviewed in detail (74). Before organ removal, animals may receive inducers such as phenobarbital or 3-methylcholanthrene for a few days, to increase the amount of drug-metabolizing enzyme. The rinsed mucosa is scraped and then homogenized in a Potter-Elvehjem homogenizer or similar device. Homogenization is a critical step. The disruption of cells must be as complete as possible without damage to the intracellular organelles. The resulting homogenate may be used as such (8, 59, 75–78) for a gross assessment of metabolism. Otherwise, subsequent two- or three-step differential centrifugation is done. Nuclei can be spun down with cell debris and brush border by a short 600–1,000 g centrifugation (78, 79). After centrifugation of the supernatant at 9,000 g (or 12,000 g), the pellet can be used as mitochondrial fraction (78) and the postmitochondrial supernatant (S9 fraction) as extract containing microsomal and cytosolic enzymes (55, 78, 80, 81). Centrifugation of the S9 fraction at 105,000 g separates the microsomal pellet, containing the endoplasmic reticulum, and the supernatant (S105 fraction), containing the soluble, non-membrane-bound enzymes (82). When only microsomes are prepared, the two-step (S9–S105) procedure (82) is used by most groups. A one-step (S105) procedure is only sufficient to work with cytosolic enzymes (83, 84). Membrane vesicles, mainly from the brush border, can be prepared from colonocytes using the method described for enterocytes (85). They are used to study the mechanisms of drug transport or the influence of absorption enhancers (86).

C. Studies of Drug Metabolism

The cytochrome P450 content of homogenates from isolated colonic cells of the guinea pig was up to 11 pmol/mg cell protein. This represents 6% of the content found in hepatocytes (65). In rat colonic microsomes, cytochrome P450 concentrations were 16 pmol/mg cell protein. They increased to 60 pmol/mg microsomal protein with phenobarbital pretreatment (87, 82). The biotransformation of procarcinogens have been studied with rat (75) and human colonic homogenates (81) and microsomes (82; references in Table 5). The role of age (79, 88), diet (75, 89), and bacterial flora (90, 91), as well as the influence of gastrointestinal hormones (92), have been assessed. The ability of human colonic epithelium to activate compounds such as benzo(a)pyrene was found to be much higher in the distal colon of patients with ulcerative colitis than in the mucosa of normal subjects. This result is relevant with respect to the increased incidence of carcinoma in such patients (76).

Homogenates and subcellular fractions have also been used to study glucuronidation (8, 47, 55, 78), sulfation (47), acetylation (59, 77), and glutathione conjugation (80). Interesting is that S9 fractions from tumorous human colon did not show the glucuronide-predominant pattern of 1-naphthol conjugation observed with isolated cells or organ cultures (47). The respective contribution of the different subcellular fractions of colonocytes to the glucuronidation of 1-naphthol (55) and morphine (78) has been elegantly illustrated and compared with liver (78) or other segments of the digestive tract (55). The colonic acetylation of 5-aminosalicylate in humans is not influenced by the acetylation phenotype (77), but it is rate-limited by acetyl CoA (59).

VI. FUTURE DIRECTIONS

This chapter has emphasized the respective contribution of various in vitro models to the knowledge of drug transport and metabolism in the colonic mucosa.

At the organ level, further developments are required, such as in vitro luminally and vascularly perfused segments to allow extrapolation to the in vivo situation. At the cellular level, a wider use of isolated cells and brush border vesicles may help to characterize the mechanisms of transcellular permeation of compounds and drug interactions at the level of absorption. The enzyme pattern of cultured cell lines of appropriate differentiation should be defined more systematically because these cells are potentially very useful to assess drug metabolism and transport in a unique situation of homogeneous monolayer of pure epithelium. More complete data about the distribution of enzymes within the colonic mucosa should be gained with immunolabeling methods in combination with systematic enzyme activity studies. Another highly interesting field would be the study of interactions between bacterial and colonic metabolism of drugs and carcinogens. Finally, in vitro models will be invaluable to attain more information about factors influencing colonic drug disposition, such as age, sex, time of day, diet, or chiral structure of drugs.

In conclusion, appropriate in vitro studies will be increasingly needed in the future as a contribution to a better understanding of colonic drug disposition in vivo.

REFERENCES

1. M. Schwenk, *Toxicol. Pathol.* 16: 138 (1988).
2. M. Schwenk, in *Intestinal Metabolism of Xenobiotics*, (A. S. Koster, E. Richter, F. Lauterbach, F. Hartmann, eds.) Gustav Fischer Verlag, Stuttgart (1989) p 155.
3. K. F. Ilett, L. B. G. Tee, P. T. Reeves, R. F. Minchin, *Pharmacol. Ther.* 46: 67 (1990).
4. M. Laitinen, J. B. Watkins, in *Gastrointestinal Toxicology* (K. Rozman, O. Hänninen, eds.) Elsevier Science, Amsterdam (1986) p 169.
5. C. F. George, *Clin. Pharmacokinet.* 6: 259 (1981).
6. K. J. W. Hartiala, M. O. Pulkkinen, P. Savola, *Nature* 201: 1036 (1964).
7. K. Kubat, O. Koldovsky, *Acta Histochem.* 33: 75 (1969).
8. O. Hänninen, A. Aitio, and K. Hartiala, *Scand. J. Gastroenterol.* 3: 461 (1968).
9. O. Hänninen, K. Alanen, K. Hartiala, *Scand. J. Gastroenterol.* 1: 152 (1966).
10. K. Hartiala, *Physiol. Rev.* 53: 496 (1973).
11. J. Kolars, P. Schmiedlin-Ren, W. Dobbins, R. Merion, S. Wrighton, P. Watkins, *FASEB* 4: A 3170 (1990).
12. W. H. Barr, S. Riegelman, *J. Pharm. Sci.* 59: 154 (1970).
13. J. T. Doluisio, N. F. Billups, L. W. Dittert, E. T. Sugita, J. V. Swintosky, *J. Pharm. Sci.* 58: 1196 (1969).
14. K. Schümann and G. J. Strugala, in *Intestinal Metabolism of Xenobiotics* (A. S. Koster, E. Richter, F. Lauterbach, F. Hartmann, eds.) Gustav Fischer Verlag, Stuttgart (1989) p 69.
15. H. G. Windmueller, A. E. Spaeth, *J. Biol. Chem.* 249:5070 (1974).
16. F. Hartmann, M. Plauth, in *Intestinal Metabolism of Xenobiotics* (A. S. Koster, E. Richter, F. Lauterbach, F. Hartmann, eds.) Gustav Fischer Verlag, Stuttgart (1989) p 81.
17. E. Scharrer, L. Stubenhofer, W. Tiemeyer, C. Bindl, *Comp. Biochem. Physiol.* 77A: 85 (1984).

18. R. Herz, D. F. Tapley, J. E. Ross, *Biochim. Biophys. Acta* 53: 273 (1961).
19. J. A. Atchison, W. E. Grizzle, D. J. Pillion, *J. Pharmacol. Exp. Ther.* 248: 567 (1989).
20. M. Tomia, M. Shiga, M. Hayashi, S. Awazu, *Pharmacol. Res.* 5: 341 (1988).
21. R. B. Sund, B. Hillestad, *Acta Pharmacol. Toxicol.* 51:377 (1982).
22. B. Hillestad, R. B. Sund, M. Buajorded, *Acta Pharmacol. Toxicol.* 51: 388 (1982).
23. R. B. Sund, S. O. Elvegård, *Pharmacology* 36: Suppl. 1, 144 (1988).
24. S. M. Rogers, D. J. Back, in *Intestinal Metabolism of Xenobiotics* (A. S. Koster, E. Richter, F. Lauterbach, F. Hartmann, eds.) Gustav Fischer Verlag, Stuttgart (1989) p 43.
25. P. Smith, C. Mirabelli, J. Fondacaro, F. Ryan, *J. Dent. Pharmacol. Res.* 5:598 (1988).
26. U. Karbach, *Gastroenterology* 96:1282 (1989).
27. G. F. Grady, R. C. Duhamel, E. W. Moore, *Gastroenterology* 59: 583 (1970).
28. S. M. Rogers, D. J. Back, L. E. Orme, *Br. J. Clin. Pharmacol.* 23: 727 (1987).
29. F. Lauterbach, *Naunyn-Schmiedebergs Arch. Pharmacol.* 297: 201 (1977).
30. G. M. Grass, S. A. Sweetana, *Pharmacol. Res.* 5: 372 (1988).
31. M. Schwenk, C. Linz, G. Rechkemmer. *Biochem. Pharmacol.* 43: 771 (1992).
32. F. Lauterbach, in *Handbook of Experimental Pharmacology*, (K. Greef, ed.) Springer Verlag, Berlin (1981) p 105.
33. F. Lauterbach, *J. Pharmacol. Exp. Ther.* 243: 1121 (1987).
34. B. Riese, K. Orth, *Naunyn-Schmiedebergs Arch. Pharmacol.* 335: R 5 (1987).
35. R. B. Sund, F. Lauterbach, *Acta Pharmacol. Toxicol.* 58: 74 (1986).
36. R. B. Sund, F. Lauterbach, *Pharmacol. Toxicol.* 60: 262 (1987).
37. M. Schorn, F. Lauterbach, *Naunyn-Schmiedebergs Arch. Pharmacol.* 332: R 11 (1986).
38. G. Sprakties, A. Karim, F. Lauterbach, *Gastroenterol. Clin. Biol.* 2: 872 (1984).
39. F. Lauterbach, M. Schorn, G. Sparkties, R. B. Sund, in *Intestinal Metabolism of Xenobiotics* (A. S. Koster, E. Richter, F. Lauterbach, F. Hartmann, eds.) Gustav Fischer Verlag, Stuttgart (1989) p 231.
40. E. Bresnick, *Cancer* 45: 1047 (1980).
41. H. P. Hoensch, F. Hartmann, *Hepatogastronterology* 28: 221 (1981).
42. C. C. Harris, B. F. Trump, R. Grafstrom, H. Autrup, *J. Cell. Biochem.* 18: 285 (1982).

43. H. Autrup, L. A. Barrett, F. E. Jackson, M. L. Jesudason, G. Stoner, P. Phelps, B. F. Trump, C. C. Harris, *Gastroenterology* 74: 1248 (1978).
44. H. Autrup, C. C. Harris, G. D. Stoner, M. L. Jesudason, B. F. Trump, *J. Natl. Cancer Inst.* 59: 351 (1977).
45. H. Autrup, C. C. Harris, B. F. Trump, A. M. Jeffrey, *Cancer Res.* 38: 3689 (1978).
46. G. M. Cohen, R. C. Grafstrom, E. M. Gibby, L. Smith, H. Autrup, C. C. Harris, *Cancer Res.* 43: 1312 (1983).
47. E. M. Gibby, G. M. Cohen, *Br. J. Cancer* 49: 645 (1984).
48. J. R. Dawson, M. Schwenk, in *Intestinal Metabolism of Xenobiotics* (A. S. Koster, E. Richter, F. Lauterbach, F. Hartmann, ed.) Gustav Fischer Verlag, Stuttgart (1989) p 21.
49. E. Hegazy, V. Lopez Del Pino, M. Schwenk, *Eur. J. Cell. Biol.* 30: 132 (1983).
50. W. E. W. Roediger, *Gastroenterology* 83: 424 (1982).
51. M. S. M. Ardawi, E. A. Newsholme, *Biochem. J.* 231: 713 (1985).
52. Y. Umesaki, T. Yajima, K. Tohyama, M. Mutai, *Pflügers Arch.* 388: 205 (1980).
53. M. Schwenk, E. Nilsson, E. K. Schmidt, *Naunyn-Schmiedebergs Arch. Pharmacol.* 341: Suppl. R 7 (1990).
54. N. Sheth-Desai, V. Lamba-Kanwal, A. Eichholz, *Jpn. J. Cancer Res.* 78: 117 (1987).
55. A. S. Koster, A. C. Frankhuijzen-Sierevogel, J. Noordhoek, *Biochem. Pharmacol.* 34: 3527 (1985).
56. M. Schwenk, M. Locher, *Biochem. Pharmacol.* 34: 697 (1985).
57. M. E. Mouelhi, M. Schwenk, *Drug Metab. Dispos.* 19: 844 (1991).
58. A. Ireland, J. D. Priddle, D. P. Jewell, *Scand. J. Gastroenterol.* Suppl. 148: 53 (1988).
59. A. Ireland, J. D. Priddle, D. P. Jewell, *Gastroenterology* 90: 1471 (1986).
60. A. Ireland, J. D. Priddle, D. P. Jewell, *Gastroenterology* 92: 1447 (1987).
61. M. Schwenk, P. Dechelotte, T. Riemenschneider, *Am. J. Physiol.* 262: G359 (1992).
62. G. Hauber, H. J. Burger, M. Schwenk, H. Remmer, *Naunyn-Schmiedebergs Arch. Pharmacol.* 322: Suppl. R 121 (1983).
63. R. A. Dale, *Br. J. Cancer* 14: 45 (1960).
64. W. E. W. Roediger, S. C. Truelove, *Gut* 20: 484 (1979).
65. H. J. Burger, G. Hauber, W. Schlote, M. Schwenk, *Am. J. Physiol.* 248: C 271 (1985).
66. V. Perret, R. Lev, W. Pigman, *Gut* 18: 382 (1977).
67. A. Herp, V. Perret, E. Suba, *Microsc. Acta* 82: 251 (1979).
68. T. P. Pretlow, T. G. Pretlow, *Birth Defects* 16: 95 (1980).
69. D. D. Maslow, E. Mayhew, A. Mittleman, *Oncology* 38: 292 (1981).
70. P. R. Gibson, E. Vna De Pol, L. E. Maxwell, A. Gabriel, W. F. Doe, *Gastroenterology* 96: 283 (1989).

71. D. P. Chopra, K. Y. Yeh, *In vitro* 17: 441 (1981).
72. M. Neutra and D. Louvard, in *Functional Epithelial Cells in Culture* (K. S. Matlin, J. D. Valentich, eds.) Alan R. Liss, New York (1989) p 363.
73. R. Muthiah, B. Seetharam, *J. Cell. Biol.* 105: 235 a (1987).
74. J. R. Dawson and R. D. Combes, in *Intestinal Metabolism of Xenobiotics* (A. S. Koster, E. Richter, F. Lauterbach, F. Hartmann, ed.) Gustav Fischer Verlag, Stuttgart (1989) p 5.
75. L. W. Wattenberg, J. L. Leong, P. J. Strand, *Cancer Res.* 22: 1120 (1962).
76. J. W. Mayhew, P. E. Lombardi, K. Fawaz, B. Goldin, A. Foster, M. Goldberg, R. Sepersky, J. M. Kellum, S. L. Gorbach, *Gastroenterology* 85: 328 (1983).
77. H. Allgayer, N. O. Ahnfeld, K. Frank, H. N. A. Söderberg, W. Kruis, G. Paumgartner, *Gastroenterology* 88: 1303 A (1985).
78. G. M. Pacifici, C. Bencini, A. Rane, *Xenobiotica* 16: 123 (1986).
79. T. F. McMahon, W. P. Beierschmitt, M. Weiner, *Cancer Letts.* 36: 273 (1987).
80. C. P. Siegers, H. Böse-Younes, E. Thies, R. Hoppenkamps, M. Younes, *J. Cancer Res. Clin. Oncol.* 107: 238 (1984).
81. C. P. Siegers, S. Schümann, E. Thies, H. Böse-Younes, M. Younes, *Cancer Letts.* 23: 39 (1984).
82. W. F. Fang, H. W. Strobel, *Arch. Biochem. Biophys.* 186: 128 (1978).
83. H. Boström, D. Brämster, H. Nordenstam, B. Wengle, *Scand. J. Gastroenterol.* 3: 369 (1968).
84. J. W. Crabtree, D. L. Dexter, J. D. Stoeckler, T. M. Savarese, L. Y. Ghoda, T. L. Rogler-Brown, P. Calabresi, R. E. Parks, *Biochem. Pharmacol.* 30: 793 (1981).
85. M. Kessler, O. Acuto, C. Storelli, H. Murer, M. Müller, G. A. Semenza, *Biochim. Biophys. Acta* 506: 136 (1978).
86. M. Tomita, M. Hayashi, T. Horie, T. Ishizawa, S. Awazu, *Pharmacol. Res.* 5: 786 (1988).
87. W. F. Fang, H. W. Strobel, *Cancer Res.* 38: 2939 (1978).
88. J. Sun, H. W. Strobel, *Exp. Gerontol.* 21: 523 (1986).
89. M. J. Wargovich, I. C. Felkner, *Nutr. Cancer* 4: 146 (1982).
90. E. C. McCoy, L. A. Petrullo, H. S. Rosenkranz, *Biochem. Biophys. Res. Commun.* 89: 859 (1979).
91. C. Larsen, E. Harboe, M. Johansen, H. P. Olesen, *Pharmacol. Res.* 6: 995 (1989).
92. W. F. Fang, H. W. Strobel, *Cancer Res.* 41: 1407 (1981).
93. R. J. Oschinsky, H. W. Strobel, *Molec. Cell. Biochem.* 75: 51 (1987).
94. L. C. Boffa, C. Bolognesi, *Muta. Res.* 173: 157 (1986).
95. S. N. Newaz, W. F. Fang, H. W. Strobel, *Cancer* 52: 794 (1983).

4
Scintigraphic Techniques in Studying Colonic Drug Absorption

John G. Hardy
Pharmaceutical Profiles Limited, Nottingham, England

I. INTRODUCTION

Gamma scintigraphy is the technique of choice for monitoring the transit of pharmaceutical dosage forms through the gastrointestinal tract. The procedure involves labeling a formulation with a gamma-emitting radionuclide. Following administration, an image of the distribution of the tracer within the body is obtained using a gamma camera. The combination of gamma scintigraphy with conventional pharmacokinetic studies facilitates the identification of sites of drug absorption.

Gamma scintigraphic studies of oral pharmaceutical dosage forms have evolved from clinical investigations of gastric emptying, in which patients consume meals labeled with nonabsorbable radiopharmaceuticals. Both liquid and solid phases of a meal can be radiolabeled using different tracers and the emptying rates measured simultaneously. The procedure is noninvasive and results in low radiation doses to the patients.

77

Although gastric emptying studies are used as an aid to diagnosis, their main clinical application is in the evaluation of responses to medical or surgical treatment (1). Studies are also undertaken to assess the efficacy of drugs intended to enhance or delay the transit of food through the stomach (2). Gamma scintigraphy is used to monitor gastrointestinal transit in patients with constipation and diarrhea (3). Such studies, however, tend to be carried out in specialized units, since they are time-consuming and difficult to schedule within busy nuclear medicine departments.

In addition to monitoring the transit of food, gamma scintigraphy can be used to investigate the behavior of pharmaceutical dosage forms within the gastrointestinal tract. Such studies are particularly relevant in the assessment of formulations designed to target specific regions of the gut, or to release active ingredients over prolonged periods.

II. RADIOLABELING

The range of radionuclides suitable for use in gamma scintigraphy is very restricted. Most gamma cameras are designed to image radionuclides emitting gamma radiations in the energy range 50–400 keV. The presence of higher-energy gamma emissions degrades the images, while alpha and beta emissions result in increased radiation doses to the subjects. The most commonly used radionuclide in nuclear medicine is technetium 99m, which emits 140 keV gamma radiation and has a half-life of 6 hr. Technetium 99m, however, is of limited use as a tracer for colonic transit studies, in which imaging periods may extend to several days. the most useful radionuclides for imaging colonic transit are indium[111] and samarium[153] (Table 1).

In studies of pharmaceutical products, it is rarely possible to radiolabel drug molecules with appropriate gamma-emitting nuclides. Usually the tracer is in the form of a compound not absorbed from the gastrointestinal tract. Thus, the images provide information on the location of a formulation within the gut and the extent of its dispersion.

Indium[111] is readily available as indium chloride solution. Indium can be adsorbed to a cation exchange resin, such as Amberlite IR-1/20(H). It remains firmly bound to the resin within the gastro-

Table 1 Radionuclides Used for Imaging Gastrointestinal Transit

Radionuclide	Half-life (hr)	Gamma ray energy (keV)	Radiation dosage (mSv/MBq)
^{113m}In	1.7	393	0.025
^{99m}Tc	6.0	140	0.017
^{171}Er	7.5	112	0.4
		296	
		308	
^{153}Sm	47	103	0.7
^{111}In	67	173	0.24
		247	

intestinal tract. The resin is usually administered as beads in a capsule (4) or as powder incorporated as an excipient in a tablet (5). Since indium^{111} is available at high specific activity, typically 2 GBq/µg, the amount of resin powder required when radiolabeling a tablet can be reduced to 1–2 mg or even less. Gastrointestinal transit studies are also undertaken using the nonabsorbable chelate diethylenetriaminepentaacetic acid (DTPA) In 111 as the radiolabel. The indium chelate is soluble in water and (^{111}In)DTPA powder can be incorporated into sustained-release tablets and used to mimic drug release. In vitro dissolution studies are required to establish the relationship between the release characteristics of the tracer and those of the drug.

In situations in which the preparation of the dosage form is a lengthy or complex procedure, or when facilities are not available for the handling of radioactive materials during the production process, neutron activation may provide a means of radiolabeling (6, 7). This procedure involves the incorporation of a nonradioactive material, such as samarium oxide enriched with the samarium^{152} isotope, which can be subsequently converted to a gamma-emitting tracer by neutron bombardment in a nuclear reactor. Samarium oxide is not absorbed from the gastrointestinal tract and during activation stable. Samarium^{152} becomes gamma-emitting samarium^{153}. When radiolabeling dosage forms by neutron activation, it is advisable to undertake preliminary in vitro studies to ensure that the irradiation process has no untoward effect on the

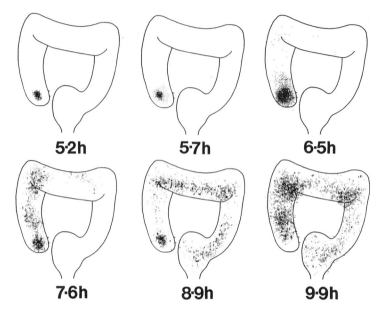

Figure 1   (a) Disintegration and dispersion of an enteric-coated, 250 mg 5-aminosalicylic acid tablet in the colon.

product. It is also important to check for unwanted radionuclides that may have been generated, for example, sodium[24], due to the presence of sodium salts in the formulation.

III. GAMMA CAMERA IMAGING

The colon can be readily identified from the images of dispersed contents (Fig. 1a). If, however, the formulation is likely to be intact on entering the colon, it is advisable to coadminister a solution radiolabeled with a different tracer to delineate the anatomy of the proximal large bowel. DTPA technetium 99m solution is often given for this purpose. Since the colon is a relatively anterior structure, radiolabeled colonic contents are best visualized on images recorded anteriorly. The morphology of the rectum and sigmoid colon, however, is such that the spreading of suppositories is best observed in images recorded in a lateral projection (Fig. 2).

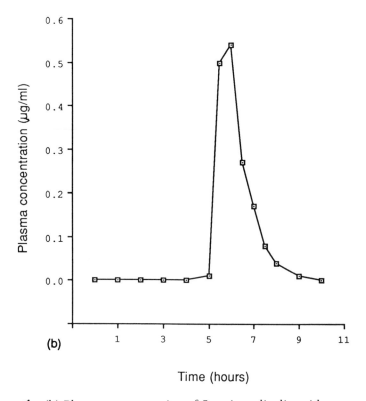

Figure 1   (b) Plasma concentration of 5-aminosalicylic acid.

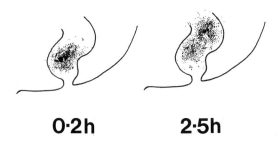

Figure 2   Spreading of a suppository.

Count rates from the same source in different regions of the colon will vary according to the extent of the attenuation of the radiation by overlying tissues. Quantification of the proportion of the tracer in each region of the colon, therefore, requires the application of attenuation corrections. The simplest and most widely adopted method is to take the geometric means of the count rates from corresponding regions of anterior and posterior images recorded at the same time (8).

IV. COLONIC TRANSIT

Gamma scintigraphy can be undertaken noninvasively and under normal physiological conditions. Therefore, it is the technique of choice for the investigation of gastrointestinal transit.

A. Oral Dosing

Formulations arrive at the ileocolonic junction about 3–4 hr after emptying from the stomach (9). The terminal ileum appears to act as a reservoir from which boluses of contents are transfered periodically into the colon (10). Transit tends to be episodic and unrelated to the consumption of food (11).

In healthy subjects, transit through the colon takes typically 20–40 hr (3, 4, 12). Most studies have reported a wide range of transit times, even when subjects have been dosed under the same conditions and with their diets well controlled. In comparison with these normal variations, differences in transit times due to the menstrual cycle (13) or the dietary restrictions of vegetarians (14) tend to be relatively minor. Most reports of intestinal transit times refer to European or North American populations. In other regions, average times may be different. For example in Southern India the average transit rates are about twice as fast (15). Most patients with common gastrointestinal complaints, such as the irritable bowel syndrome (16) and quiescent inflammatory bowel disease (17), exhibit transit times within the wide normal range. Transit through the colon is likely to be prolonged, to about twice that in healthy subjects, in patients with constipation (3, 18). Conversely, transit times are relatively short in patients with diarrhea (18).

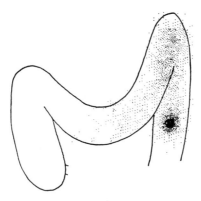

Figure 3 Sustained-release tablet in the descending colon ahead of the released tracer.

Gamma scintigraphic studies have shown the ascending and transverse sections of the colon to be the main sites for the accumulation of materials within the gastrointestinal tract (12). About half of the contents of the large bowel are found in the cecum and ascending colon (19). Transit through the proximal colon comprises periods of relative stasis, which may persist for several hours, interspersed with episodes of propulsion. The increased colonic motility resulting from eating tends to cause segmental pressure changes within the colon rather than aboral propulsion of the contents (20). Prolonged residence tends to occur at the hepatic and splenic flexures. Occasionally, aboral propulsion is followed quickly by retrograde movement (16, 20). Relatively little material is retained in the descending colon or the rectosigmoid region (12). Following evacuation of the bulk of a dispersed tracer, residual radioactivity is frequently detected in the cecum. Residence times in the proximal colon are dependent on particle size (4). Large units, such as capsules and tablets, pass through the colon faster than small pellets, solutions, or released contents (Fig. 3).

B. Rectal Dosing

The extent of dispersion of rectally administered formulations is influenced principally by the volume infused (21). In general,

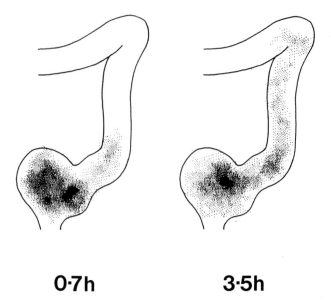

0·7h   **3·5h**

Figure 4   Spreading of a high-bulk 5-aminosalicylic acid foam enema in a patient with ulcerative colitis.

enema solutions of 100 ml will spread throughout the sigmoid and descending colon and occasionally into the ascending colon (22). Similar spreading can be achieved with foam enemas, if they are of sufficient bulk (Fig. 4). Enema dispersion tends to occur over the 2–3 hr following dosing. Subsequently, aboral movement of the colonic contents returns the preparation toward the rectum. As shown in Figure 2, the spreading of suppositories is confined mainly to the rectum (23).

V. GAMMA SCINTIGRAPHY AND DRUG ABSORPTION

Gamma scintigraphy allows monitoring of position, movement, and extent of dispersion of a drug preparation within the gastrointestinal tract under normal physiological conditions. Accurate quantification of the proportions of the preparation in different regions can be obtained. Thus, relating the location of the tracer in the gamma camera images to pharmacokinetic data derived from blood or urine assays provides information about the sites and extents of drug absorption (Fig. 1). However, it is usually an excipient of the formulation, rather than the drug, that is radiolabeled. In general, a nonabsorbable tracer will be chosen. Therefore, radioactivity may be detected in the intestines even if all the drug has been absorbed.

Two approaches to oral drug delivery can be conveniently applied in the investigation of colonic absorption using gamma scintigraphy: bolus release and sustained release. Bolus release requires that the drug will be contained within the dosage form until the preparation has entered the colon. This can be achieved by the application of enteric coatings to tablets, particulates, or capsules (24). Such a coating should be resistant to dissolution in the acidic environment in the stomach. It should dissolve gradually as the preparation passes through the small intestine, where the pH is typically 6.6–7.5 (25). This can be readily achieved with methacrylate esters that have dissolution thresholds in the range pH 6.5–7.0. The thickness of the coating should be such that drug release is prevented for about 4 hr after leaving the stomach.

Variations in arrival times of preparations in the colon can be minimized by dosing after an overnight fast (26). This is particularly pertinent when dosing with timed-release formulations without enteric coatings. The presence of food in the stomach may delay gastric emptying for several hours, resulting in drug release before the arrival of such preparations in the colon. Within a fasted subject, however, gastric emptying of a tablet will usually occur within 1 hr of dosing. Sustained-release preparations may take the form of polymeric matrix tablets or osmotic pumps that can deliver drugs along with radioactive tracer at constant rates for periods of 10 hr or more (27). Such systems can be enteric coated to prevent drug release in the stomach and small intestine.

The main disadvantage in the use of enteric-coated or timed-release preparations in the investigation of drug absorption from the colon is the intersubject variation in arrival times at the cecum. Thus, in some subjects the onset of drug release may occur in the small intestine, while in others the formulation may pass through the ascending colon intact. The application of gamma scintigraphy, however, provides a convenient means of monitoring the integrity of the formulation. The frequent blood sampling usually required during the early period following drug release can be delayed until dispersion of the tracer has been detected in the gamma camera images.

Drug release at a precise location in the colon can be achieved in a reproducible manner by the use of sophisticated delivery devices such as the high-frequency capsule, which releases the drug in response to an externally transmitted radio signal (28). Gamma scintigraphy can be used to confirm the location of the device prior to actuation, and to monitor dispersion of the released contents.

VI. CONCLUSION

The role of the colon in drug absorption is assuming an increasing importance in the development of drug delivery systems. Drug delivery from sustained-release products may occur primarily within the large bowel. The colon may also provide a route of absorption for drug molecules that would be degraded in the more proximal regions of the gastrointestinal tract.

A thorough understanding of the factors affecting drug absorption from the colon is required in order to exploit fully the potential of this region for systemic drug delivery. Gamma scintigraphy enables drug formulations to be monitored noninvasively, under normal physiological conditions and in the presence of pathologic changes. Thus, information is provided regarding the location, extent of dispersion, and rate of transit of a preparation, which aids in the interpretation of conventional pharmacokinetic data obtained during such studies.

REFERENCES

1. L. S. Malmud, R. A. Vitti, *J. Nucl. Med.* 31: 1499 (1990).
2. H. Minami, R. W. McCallum, *Gastroenterology* 86: 1592, (1984).
3. R. G. McLean, R. C. Smart, D. Gaston-Parry, S. Barbagallo, J. Baker, N. R. Lyons, C. E. Bruck, D. W. King, D. Z. Lubowski, N. A. Talley, *J. Nucl. Med.* 31: 985 (1990).
4. J. G. Hardy, C. G. Wilson, E. Wood, *J. Pharm. Pharmacol.* 37: 874 (1985).
5. J. G. Hardy, D. F. Evans, I. Zaki, A. G. Clark, H. H. Tonnesen, O. N. Gamst, *Int. J. Pharmacol.* 37: 245 (1987).

6. A. Parr, M. Jay, *Pharmacol. Res.* 4: 524 (1987).
7. J. G. Hardy, G. L. Lamont, D. F. Evans, A. K. Haga, O. N. Gamst, *Aliment. Pharmacol. Ther.* in press (1991).
8. J. G. Hardy, A. C. Perkins, *Nucl. Med. Commun.* 6: 217 (1985).
9. S. S. Davis, J. G. Hardy, J. W. Fara, *Gut* 27: 886 (1986).
10. M. Camilleri, L. J. Colement, S. F. Phillips, M. L. Brown, G. M. Thomforde, N. Chapman, A. R. Zinsmeister, *Am. J. Physiol.* 257: G 284 (1989).
11. M. J. Mundy, C. G. Wilson, J. G. Hardy, *Nucl. Med. Commun.* 10: 45 (1989).
12. M. Proano, M. Camilleri, S. F. Phillips, M. L. Brown, G. M. Thomforde, *Am. J. Physiol.* 258: G 856 (1990).
13. M. A. Kamm, M. J. G. Farthing, J. E. Lennard-Jones, Gut 30: 605 (1989).
14. G. J. Davies, M. Crowder, B. Reid, J. W. T. Dickerson, Gut 27: 164 (1986).
15. V. Jayanthi, A. Chacko, I. Karim Gani, V. I. Mathan, Gut 30: 35 (1989).
16. J. G. Hardy, E. Wood, A. G. Clark, J. R. Reynolds, *Eur. J. Nucl. Med.* 11: 393 (1986).
17. J. G. Hardy, J. N. C. Healey, J. R. Reynolds, *Aliment. Pharmacol. Ther.* 1: 273 (1987).
18. P. A. Cann, N. W. Read, C. Brown, N. Hobson, C. D. Holdsworth, *Gut* 24: 405 (1983).
19. J. H. Cummings, E. W. Pomare, W. J. Branch, C. P. E. Naylor, G. T. Macfarlane, Gut 28: 1221 (1987).
20. D. J. Holdstock, J. J. Misiewicz, T. Smith, E. N. Rowlands, Gut 11: 99 (1970).
21. J. G. Hardy, S. W. Lee, A. G. Clark, J. R. Reynolds, *Int. J. Pharmacol.* 31: 151 (1986).
22. E. Wood, C. G. Wilson, J. G. Hardy, *Int. J. Pharmacol.* 25: 191 (1985).
23. J. G. Hardy, L. C. Feely, E. Wood, S. S. Davis, *Int. J. Pharmacol.* 36: 103 (1987).
24. J. N. C. Healey, in *Drug Delivery to the Gastrointestinal Tract* (J .G. Hardy, S. S. Davis, C. G. Wilson, eds.) Ellis Horwood, Chichester (1989) p 83.
25. D. F. Evans, G. Pye, R. Bramley, A. G. Clark, T. J. Dyson, J. D. Hardcastle, Gut 29: 1035 (1988).
26. J. G. Hardy, J. N. C. Healey, S. W. Lee, J. R. Reynolds, *Aliment. Pharmacol. Ther.* 1: 209 (1987).
27. A. Chacko, K. F. Szaz, J. Howard, J. H. Cummings, *Gut* 31: 106 (1990).
28. A. H. Staib, D. Loew, S. Harder, E. H. Graul, R. Pfab, *Eur. J. Clin. Pharmacol.* 30: 691 (1986).

5

Other Methods in Studying Colonic Drug Absorption

Karl-Heinz Antonin
Ciba-Geigy, Tübingen, Germany

I. INTRODUCTION

Most drugs are commonly administered orally as solid dosage forms. Most orally administered drugs are designed to be absorbed from the upper gastrointestinal tract. In recent years, efforts have been made to control the rate and amount of drug released, and the site of delivery, and to understand better the uptake mechanisms. Sustained-release dosage formulations were developed to provide convenient once-daily administration and to prevent toxic effects. These new formulations deliver drugs at a constant rate along the entire gut, avoiding peak blood concentrations and related adverse effects.

A special type of sustained-release formulation is the osmotic drug delivery system (OROS). Water from the gut diffuses through a semipermeable membrane, dissolves the drug, and (in the push–pull device) simultaneously an osmotic agent contained in the core of the system. The resulting solution is expelled through a tiny laser-drilled exit hole (1–3).

In humans mean transit time from mouth to cecum is 3–7 hr (4, 5) while that through the entire gut exceeds 20 hr (6–9). If the releasing time exceeds 7 hr, part of the dose will have to be absorbed by the colonic mucosa. Bioavailability might be restricted by the inability of the colon to absorb the drug or by bacterial degradation in the colon. Therefore, development of new dosage formulations that release drug slowly along the entire gastrointestinal tract calls for systematic studies of the absorptive capacity of all parts of the gastrointestinal tract (10, 11).

The absorptive function of the human colon has until now been studied mainly in conjunction with the transport of water, electrolytes, and ammonia across the mucosa (12). Relatively little attention has been given to drug absorption in this segment of the human gut.

Resulting from biotechnological advances, peptide drugs will become more readily available at reasonable and affordable costs in the quantities needed for drug therapy. Many hormone deficiency diseases call for long-term treatment. Proteins or peptides have to be administered subcutaneously, intramuscularly, or intravenously. Oral administration is prevented by poor absorption because of the large molecular size and digestion in the upper gastrointestinal tract. Improvement in compliance and better disease control support the search for alternative delivery routes. In recent years, it has been shown in animal experiments that proteins or peptides, once they have passed the stomach and the intestine, are most likely to be absorbed, particularly in the colon (13–15). With new oral drug delivery systems it may be possible to release proteins or peptides within a desired region of the human colon where they may be absorbed. Therefore, it is important to explore whether and to what extent drugs or peptide hormones will be absorbed by the colonic mucosa.

Several techniques, including indirect methods, colonic intubation, and, more recently, remote-controlled drug delivery devices, are available to study colonic drug absorption in humans (10, 11, 16, 17). However, so far few data are available on colonic drug absorption in humans. This chapter will review, summarize, and assess methods not described in Chapters 4 and 6. It will focus only on techniques used in human research.

II. METHODS

A. Indirect Methods

Indirect methods for the measurement of drug absorption from the lower part of the gastrointestinal tract are based on the evaluation of pharmacokinetic data. Oral slow-release preparations are used or formulations designed to be degraded by micro-organisms in the colon. From fraction absorbed time profiles after sustained-release and fast-release formulations the retardation of drug liberation can be calculated. Absorption from the colon can be estimated if the mouth-to-cecum transit time is known. It can be measured simultaneously by the lactulose breath test or by measurement of the appearance of sulfapyridine in blood or saliva after intake of salicylazosulfapyridine (18). With this indirect method the fraction of drugs absorbed from the large bowel was assessed after administration of an OROS or a sustained-release theophylline formulation (19–21).

The human gastrointestinal tract has a complex ecosystem of micro-organisms that plays an important role in the metabolism of nutrients and drugs. In the colon, bacteria metabolize a large spectrum of different substrates. Specific drugs or drug formulations escaping small bowel digestion come intact into the colon. Here, they are degraded by micro-organisms and then absorbed. Experience has been gained with salicylazosulfapyridine, the azo prodrug of 5-amino salicylic acid. As it reaches the colon after oral administration, the diazo-binding is cleaved by colonic bacteria into sulfapyridine and 5-amino salicylic acid. Sulfapyridine is absorbed and can be measured in plasma or saliva. The plasma concentration time curve (Fig. 1) shows continuous absorption from the colon. Absolute bioavailability is 70%, compared with intravenous administration of sulfapyridine (10).

All indirect methods can only give a general impression of the conditions of absorption. They do not show how the drug is removed from the gastrointestinal tract and where absorption takes

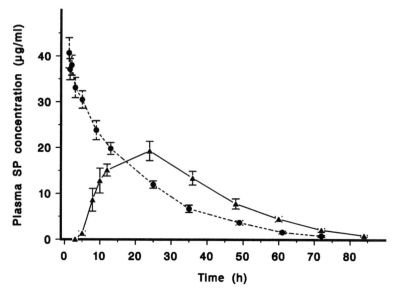

Figure 1 Sulfapyridine (SP) plasma concentration–time curves of six healthy subjects after intravenous sulfapyridine (●) and after oral salazo-sulfapyridine (Colo-Pleon) administration (▲) (mean ± SEM).

place. It also is not taken into account that a sustained-release formulation is embedded in almost solid feces in the distal part of the colon and that this may prevent further absorption.

B. Retrograde Enema

Topically acting or poorly absorbed drugs such as 5-amino salicylic acid or prednisolone metasulfobenzoate are given by rectal enemas for the treatment of distal inflammatory bowel diseases. In normal volunteers (22) and in patients (23) it has been shown that radiolabeled enema solutions, having volumes of 100 ml or more, spread beyond the sigmoid colon to the transverse colon within 1–4 hr after application. It seems possible to apply this method for the investigation of colonic drug absorption.

Neomycin administered by enema and orally was measured in urine and serum. Abdominal radiographs taken 30 min after rectal administration of a mixture of barium sulfate and neomycin showed that the enema had reached the splenic flexure. Absorption after enema was like that after oral administration in normal subjects and in several groups of patients. Increasing the enema retention time did not increase neomycin absorption (24).

C. Intubation Techniques

Drug absorption in humans can be studied by oral intubation techniques. Such techniques were originally used for exploring the absorption of water, ions, and nutrients. Recently, they have been used for the investigation of the gastrointestinal absorption of drugs in humans (10, 16, 17).

For the examination of segmental absorption, multichannel tubes are used. Absorption rates are estimated from the rate of disappearance of drug from the gut lumen. Nonabsorbable markers such as polyethylene glycol (PEG) are used to allow for correction of changing volumes resulting from fluid fluxes. Thin flexible tubes are swallowed by the subject and placed in different parts of the gastrointestinal tract. The position is controlled by radiography. With inflated occlusive balloons attached to the tubes, intestinal segments can be isolated. Substances and markers are continuously infused and luminal content is collected at various distal sites and intervals (16, 17). To study drug absorption from the colon, the subject swallows a tube. Its distal end is positioned into the cecum under radiologic control. The drug is administered through the tube and the rate of absorption is calculated from plasma concentration profiles.

Using such intubation techniques, the absorption of the two β-adrenoceptor antagonists metoprolol and oxprenolol from different sites of the gastrointestinal tract has been studied (17, 25–29). Plasma concentrations after administration of metoprolol (Fig. 2) or oxprenolol into the jejunum, ileum, and the proximal part of the colon were similar.

The application of these intubation techniques with calculation of the disappearance of substance from the intestinal lumen is based on the assumption that absorption is the sole mechanism of luminal clearance. It is assumed that no other processes occur, such as intestinal metabolism, enzymatic or bacterial breakdown, or binding of compounds to the mucosal surface. Intubation tech-

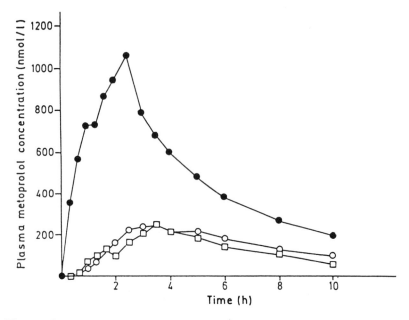

Figure 2   Mean plasma concentration–time curves of metoprolol after intravenous (●), intrajejunal (□), and intracolonic (○) perfusion for 2.5 hr (n = 6) (from ref. 27).

niques are complicated, inconvenient for volunteers, costly, and time consuming. Therefore, they cannot be used extensively.

D. Remote-Controlled Drug Delivery Devices

1. High-Frequency Capsule

A high-frequency capsule has been developed to study absorption characteristics in humans (30, 31). It allows the study of extent and rate of drug absorption from all sites of the gastrointestinal tract. Figure 3 shows a diagram of the capsule. A smooth plastic capsule (12 × 28 mm) contains a small latex balloon (1 ml) filled with drug and the release mechanism. Solutions, suspensions, and solids (especially micropellets) are suitable as filling. The capsule is swallowed with a small oral dose of contrast medium to allow radio-

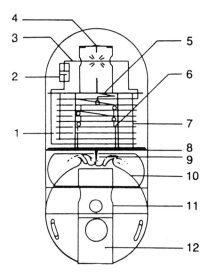

Figure 3 Schematic diagram of the high-frequency capsule. 1, Oscillating circuit; 2, capacitor; 3, heating wire; 4, nylon thread; 5, spring; 6, plunger; 7, cylinder liner; 8, separation wall; 9, small steel needle; 10, latex container; 11, holder for 10; 12, plug (from ref. 31).

logic localization. The release mechanism is triggered with a short impulse from a 27 MHz high-frequency generator and transduced by a specially designed ring antenna to release the drug into a defined part of the gastrointestinal tract. Drug release at different parts within the same subject allows for comparison of absorption parameters. From plasma concentration profiles, the rate of absorption, absorption half-life, and mean residence time can be calculated.

In recent years, the high-frequency capsule was used in healthy volunteers to study the absorption of several drugs from different segments of the gastrointestinal tract (Table 1) (31–36). Comparison of the area under the curve shows that the colonic absorption of these drugs differs markedly. Theophylline, isosorbide-5-mononitrate, metoprolol, and nitrendipine showed no difference in extent and rate of absorption from the colon compared with oral administration. In contrast, the extent of absorption from the colon was much lower for furosemide and ciprofloxacin, showing a so-

Table 1   Bioavailability (BAV) of Drugs Measured After Intracolonic Administration with the High-frequency Capsule

Drug	Comparison	Colon BAV	n	Reference
Furosemide	p.o.	3%	3	32
Isosorbide-5-MN	i.g.	91%	2	33
Theophylline	i.g.	86%	3	34
	p.o.	91%	7	35
Ciprofloxacin	p.o.	7%	4	36
Metoprolol	i.g.	109%	10	31
Nitrendipine	p.o.	72%	6	35
Nisoldipine	p.o.	311%	4	35

p.o., orally; i.g., intragastrically.

called "absorption window" (37) for these drugs. The Ca^{2+} antagonist nisoldipine showed a threefold increase of bioavailability after colonic application compared to an oral dose. The bioavailability of dilazep hydrochloride *(rINNM)* after colonic release was nearly 15-fold greater than after oral administration, suggesting a reduced prehepatic first-pass loss.

2. Telemetric Capsule

Recently, a new telemetric multifunctional capsule has been developed (38) (Fig. 4A). The device consists of a plastic cylinder (39 × 11 mm; weight 3.5 g). It is separated into two parts. The main element contains a lithium battery, a radio transmitter able to transmit signals over a distance of 10 m, a location detector, and a magnetic switch. The location detector allows continuous data collection during the passage through the gastrointestinal tract. The second element is a reservoir that can be fixed to the front of the capsule, permitting remote control release or pick up of 1 ml of a liquid. The release tip (Fig. 4B) is a latex reservoir with 1 ml capacity. The reservoir is filled with the drug using a syringe. The magnetic switch in the main element allows remote-control release. The release mechanism is triggered by an external permanent magnet that must be brought close to the capsule. This new device allows detailed investigation of intestinal absorption from

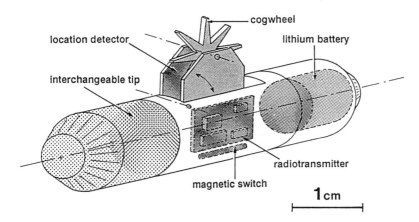

Figure 4 A. Schematic diagram of the telemetric capsule. Rotations of the cogwheel induced by the progression of the capsule in the small bowel are transmitted by the radiotransmitter. The interchangeable tip is remotely controlled by the magnetic switch. A lithium battery supplies the device for 24 hr (from ref. 38).

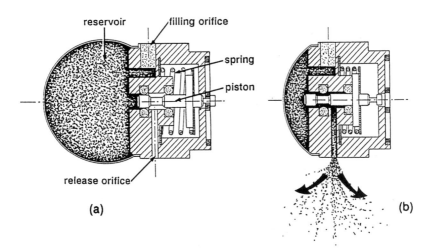

Figure 4 B. Schematic diagram of the release tip. (a) A piston keeps the latex reservoir closed, which is filled with substance to be released. (b) After remote control, the piston moves on, allowing the release of the substance through the release orifice (from ref. 38).

different segments of the gastrointestinal tract under physiological conditions.

E. Colonoscopy

To evaluate the absorptive capacity of the colonic mucosa, colonoscopy has been used in healthy volunteers or patients. With the 180 cm long colonoscope, the cecum and even the terminal ileum can be reached easily. Through instrumental channels, dissolved drugs can be placed into distinct regions of the colon under visual and fluoroscopic control. From plasma concentration time profiles the rate of absorption can be calculated. Done by an experienced endoscopist, colonoscopy is a safe and quick method. Complications are very rare. The most dangerous (perforation and heavy bleeding) occur in less than 0.3% of procedures (39) and most often in patients with a colon disease such as inflammation, diverticulosis, or malignancy. However, when using colonoscopy for studying drug absorption in healthy volunteers, these risks are negligible. A well-trained endoscopist can do a complete examination, identifying the parts of the colon to the ileocecal valve and placing the drug into distinct areas, within 5–20 min.

The large intestine must be well cleaned before endoscopy. This method has been criticized because the absorption would be assessed under nonphysiological conditions. It has also been argued that the laxatives used may produce superficial inflammation of the colonic mucosa, thus changing the absorptive properties. However, from clinical experience and our own unpublished results it is well-known that biopsies obtained from a normal-appearing colonic mucosa during colonoscopy in the cleaned gut show normal histologic findings. Also, pretreatment with cleansing enemas was shown not to influence the absorption of neomycin given by enema (24).

In recent years, colonoscopy was used in healthy volunteers and in patients to study the absorption of several drugs from different segments of the colon (Figs. 5, 6; Table 2) (10, 40–51).

In our own studies with the β-adrenoceptor antagonist oxprenolol and the nonsteroidal anti-inflammatory agent diclofenac (42, 43, 51), the site of application (cecum or left flexure) did not influence absorption. This is interesting because regional differences in the absorptive capacity of the human colon for salt and water are

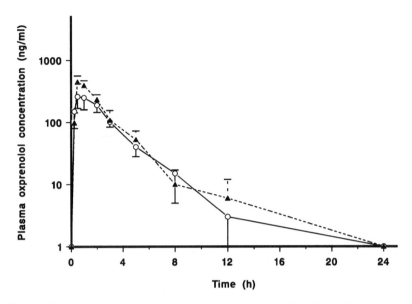

Figure 5   Plasma concentration–time curves of six healthy subjects after oral (▲) and colonic (○) application of 80 mg oxprenolol (mean ± SEM).

well-known. The ascending colon deals with large volumes of ileal effluent and absorbs salt avidly. Its mucosa is considered to be better permeable than that of the descending colon, which can maintain large concentration gradients generated across the mucosa (52).

This new indication for colonoscopy proved to be simple and seems suited for phase I studies with new compounds or new drug formulations. Dissolved drug can be placed into distinct areas under visual and fluoroscopic control. Colonoscopy allows for the quantitation of absolute or relative colonic bioavailability of a drug. With good endoscopic technique performed by a trained endoscopist and with the full flexible colonoscopes, the examination is well tolerated by subjects. Minor discomfort may result from the inflated air or if the mesentery of the bowel is stretched because of loops formed by the colonoscope. Most loops occur in a mobile redundant sigmoid colon. Disadvantages are minor discomfort to the volunteers, the need for a clean colon, and exposure to x-rays. Colonscopy should only be done by skilled endoscopists.

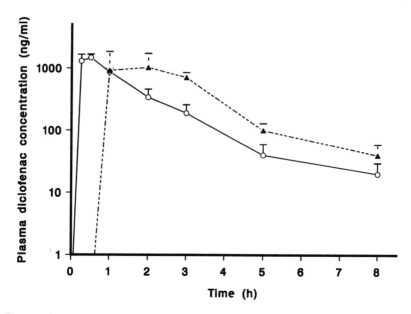

Figure 6 Plasma concentration–time curves of six healthy subjects after oral (▲) and colonic (O) application of 100 mg diclofenac sodium (mean ± SEM).

F. Artificial Stomas

Drug absorption by the colonic mucosa can also be explored in patients with intestinal stomas if no stenosis is present. Some clinical conditions require surgical creation of stomas as curative or palliative therapy. Intestinal stomas are used for the treatment of malignant or benign diseases of the gastrointestinal tract. They can be constructed temporarily or permanently as loop or end stomas. Stomas can be used in two ways to study drug absorption (10, 51). From patients with end stoma it is possible to sample intestinal content of the upper parts of the gastrointestinal tract. After oral drug intake, the nonabsorbed part of the dose can be recovered via the stoma. Measurement of drug concentrations in the feces provides an indication of the extent of absorption between mouth and stoma. The results from patients with end stomas of different parts of the gastrointestinal tract may give hints of the site and the extent of absorption. Furthermore, this method may also be

Table 2   Bioavailability (BAV) of Drugs Measured After Intracolonic
Administration with the Colonoscope

Drug	Comparison	Colon BAV	Subjects	Reference
Digoxin	p.o. 2 groups	67%	2 patients	40
Digoxin	p.o. 2 groups	85%	8 patients	41
Digoxin	p.o. 2 groups	9%	4 patients/ Colitis	41
Oxprenolol	p.o. crossover	82%	6 volunteers	42
Diclofenac	p.o. crossover	83%	6 volunteers	43
Glibenclamide	i.g. 2 groups	77%	6 patients	44
Piretanide	i.g. 2 groups	27%	6 patients	45
Theophylline	i.v. cross-over	54%	8 volunteers	46
Nicardipine		low	6 patients	47
Ursodeoxycholic acid	p.o. 2 groups	80%	patients	48
	p.o. crossover	61%	5 volunteers	49
Acrivastine	p.o. crossover	18%	6 volunteers	50
Human calcitonin	i.v. crossover	0.12%	8 volunteers	Own results, unpublished

p.o., orally; i.g., intragastrically.

valuable for gaining information on effective bactericidal intracolonic concentrations after oral antimicrobial therapy (10, 51, 53)
(Fig. 7) or on therapeutic intracolonic drug concentrations after
new oral formulations of 5-amino salicylic acid (54).

In patients with a loop stoma, drugs or special drug formulations
can be applied to the distal part of the stoma. By measuring drug
concentrations in body fluids, the absorptive capacity between
stoma and anus can be assessed. The results gained after application of an oxprenolol OROS into a stoma confirm qualitatively the
data obtained after intracolonic administration of oxprenolol in
healthy young volunteers during colonoscopy (Fig. 8). Bioavailability of morphine after application in a colostomy compared to rectal
administration of a suppository was low (55). However, absorption of theophylline compared to data after oral administration was about 60% (56). With this technique it could be shown
that theophylline is well absorbed by the colonic mucosa, thus con-
firming results using the high-frequency capsule or colonoscopy
(see Table 1 and 2).

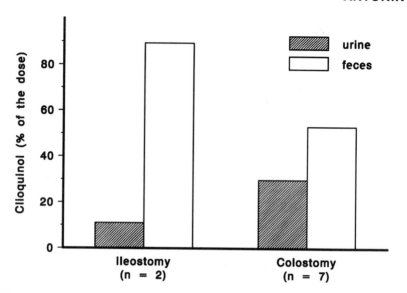

Figure 7   Cumulative excretion of clioquinol in urine and intestinal contents of patients with stoma after oral administration of 250 mg.

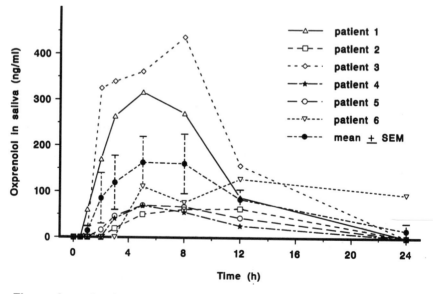

Figure 8   Individual oxprenolol saliva concentration–time curves of six patients after application of one oxprenolol OROS dose into the distal part of colonic stoma.

Investigations in patients with artificial stomas are not invasive or uncomfortable. However, the conditions of absorption are variable and the handling of gut contents is unpleasant. It must be carefully monitored so that the dose administered into the stoma is retained in the distal part. The absorptive capacity of only one part of the gastrointestinal tract can be assessed.

III. CONCLUSION

The development of dosage formulations with extended drug release requires basic knowledge of the absorption of compounds in the entire gastrointestinal tract. All techniques mentioned can be used to study drug absorption by the colonic mucosa in humans, but all have advantages and disadvantages.

Intubation techniques are highly sophisticated and permit exact characterization of the absorption processes in all segments of the gastrointestinal tract. However, intubation techniques are complicated, costly, time-consuming and a burden for the study subject. Therefore, they cannot be used routinely.

High-frequency capsules may become the most useful method for the investigation of rate and extent of drug absorption from any region of the gastrointestinal tract. However, at present this technique is not generally available. Limits of this method include the need for repeated x-ray controls, the sometimes difficult placement of a drug into the capsule, and the fact that the capsule can be used only once.

Colonoscopic application of drugs in healthy volunteers seems to be a simple and suitable method for studying drug absorption from the colon. This method, done by a skilled endoscopist, allows for the quantitative measurement of the relative colonic bioavailability of a drug. It causes minor discomfort to the volunteers, requires a clean colon, and radiologic equipment. Investigations in patients with artificial stomas are without discomfort to the patients, but they can give only qualitative indications of the absorptive capacity of the part of the gastrointestinal tract explored. Measuring saliva concentrations of drugs in patients with a stoma can prevent needless blood loss.

Occasionally, in some volunteers a higher relative bioavailability may be found after administration during colonoscopy or into stomas. This may result from colonic effluents into the rectum. Rectal absorption is known partly to bypass the liver (57). The reduced hepatic first-pass effect could explain the higher amount of drug available systemically.

In studies in humans, consideration should first be given to the magnitude of the expected pharmacologic effect. Based on this, a rational choice of methods can be made. However, it has to be acknowledged that often the most sensitive and accurate measurements are difficult to do.

REFERENCES

1. F. Theeuwes, in *Drug Absorption: Proceedings of the Edinburgh International Conference* (L. F. Prescott, W. S. Nimmo, eds.) ADIS Press, Hong Kong (1981) p 157.
2. J. Urquhart, *Drugs* 23: 207 (1982).
3. S. S. Davis, J. W. Fara, in *Drug Delivery to the Gastrointestinal Tract* (J. G. Hardy, S. S. Davis, C. G. Wilson, eds.) Ellis Horwood, Chichester (1989) p 97.
4. K. H. Antonin, P. R. Bieck, C. Schick, B. Steidel, *Z. Gastroenterol.*, 20: 554 (1982).
5. S. S. Davis, J. G. Hardy, J. W. Fara, *Gut* 27: 886 (1986).
6. J. H. Cummings, H. S. Wiggins *Gut* 17: 219 (1976).
7. V. A. John, P. A. Shotton, J. Moppert, W. Theobald, *Br. J. Clin. Pharmacol.* 19: 203S (1985).
8. N. W. Read, C. A. Miles, D. Fisher, A. M. Holgate, N. D. Kime, M. A. Mitchell, A. M. Reeve, T. B. Roche, M. Walker, *Gastroenterology* 79: 1276 (1980).
9. S. S. Davis, in *Novel Drug Delivery and its Therapeutic Application* (L. F. Prescott, W. S. Nimmo, eds.) John Wiley, Chichester (1989) p 89.
10. P. R. Bieck, in *Drug Delivery to the Gastrointestinal Tract* (J. G. Hardy, S. S. Davis, C. G. Wilson, eds.) Ellis Horwood, Chichester (1989) p 147.
11. J. W. Fara, in *Novel Drug Delivery and its Therapeutic Application* (L. F. Prescott, W. S. Nimmo, eds.), John Wiley, Chichester (1989) p 103.
12. R. Shields, in *Scientific Basis of Gastroenterology* (H. L. Duthie, K. G. Wormsley, eds.), Churchill Livingstone, Edinburgh/London/New York (1979) p 398.
13. E. Tomlinson, in *Novel Drug Delivery and its Therapeutic Application* (L. F. Prescott, W. S. Nimmo, eds.) John Wiley, Chichester (1989) p 245.
14. W. A. Ritschel, *Bull. Tech. Gattefosse Rep.* 83: 7 (1990).

15. P. Goddard, *Adv. Drug Delivery Rev.* 6: 103 (1991).
16. J. Hirtz, *Br. J. Clin. Pharmacol.* 19: 77S (1985).
17. J. Hirtz, in *Drug Absorption at Different Regions of the Human Gastro-intestinal Tract: Methods of Investigation and Results* (N. Rietbrock, B. G. Woodcock, A. H. Staib, D. Loew, eds.) Friedr. Vieweg & Sohn, Braunschweig/Wiesbaden (1987) p 3.
18. K. H. Antonin, P. R. Bieck, in *Early Phase Drug Evaluation in Man* (J. O'Grady, O. I. Linet, eds.) MacMillan Press, Houndmills (1990) p 457.
19. C. Kehe, K. Wick, S. McCarville, K. Rhode, B. C. Meakin, D. Porter, L. Harrison, *Eur. J. Clin. Pharmacol.* 40: 319 (1991).
20. B. C. Lippold, in *Drug Absorption at Different Regions of the Human Gastro-intestinal Tract: Methods of Investigation and Results* (N. Rietbrock, B. G. Woodcock, A. H. Staib, D. Loew, eds.) Friedr. Vieweg & Sohn, Braunschweig/Wiesbaden (1987) p 63.
21. De K. Sommers, E. C. Meyer, M. van Wyk, J. Moncrieff, *Eur. J. Clin. Pharmacol.* 38: 171 (1990).
22. J. G. Hardy, E. Wood, A. G. Clark, J. R. Reynolds, *Eur. J. Nucl. Med.* 12: 176 (1986).
23. M. M. C. Tiel-van Buul, C. J. J. Mulder, E. A. van Royen, E. H. H. Wiltink, G. N. J. Tytgat, *Clin. Pharmacokinet.* 20: 247 (1991).
24. K. J. Breen, R. E. Bryant, J. D. Levinson, S. Schenker, *Ann. Intern. Med.* 76: 211 (1972).
25. G. Jobin, A. Cortot, J. Godbillon, M. Duval, J. P. Schoeller, J. Hirtz, J. J. Bernier, *Br. J. Clin. Pharmacol.* 19: 97S (1985).
26. N. Vidon, D. Evard, J. Godbillon, M. Rongier, M. Duval, J. P. Schoeller, J. J. Bernier, J. Hirtz, *Br. J. Clin. Pharmacol.* 19: 107S (1985).
27. J. Godbillon, D. Evard, N. Vidon, M. Duval, J. P. Schoeller, J. J. Bernier, J. Hirtz, *Br. J. Clin. Pharmacol.* 19: 113S (1985).
28. N. Vidon, R. Palma, J. Godbillon, C. Franchisseur, G. Gosset, J. J. Bernier, J. Hirtz, *Clin. Pharmacol.* 26: 611 (1986).
29. J. Godbillon, N. Vidon, R. Palma, A. Pfeiffer, C. Franchisseur, M. Bovet, G. Gosset, J. J. Bernier, J. Hirtz, *Br. J. Clin. Pharmacol.* 24: 335 (1987).
30. A. Zimmer, W. Roth, B. Hugemann, W. Spieth, F. W. Koss, A novel method to study drug absorption. Evaluation of the sites of absorption with a capsule for wireless controlled drug liberation in the GI-tract, in *Proceedings of First European Congress on Biopharmaceuticals and Pharmacokinetics*, vol. 2, (J. M. Aiache, J. Hirtz, eds.) Paris, Technique and Documentation (1981) p 211.
31. O. Schuster, B. Hugemann, in *Drug Absorption at Different Regions of the Human Gastro-intestinal Tract: Methods of Investigation and Results* (N. Rietbrock, B. G. Woodcock, A. H. Staib, D. Loew, eds.), Friedr. Vieweg & Sohn, Braunschweig/Wiesbaden (1987) p 28.

32. E. H. Graul, D. Loew, O. Schuster, *Therapiewoche* 35: 4277 (1985).
33. H. Laufen, A. Wildfeuer, in *Drug Absorption at Different Regions of the Human Gastro-intestinal Tract: Methods of Investigation and Results* (N. Rietbrock, B. G. Woodcock, A. H. Staib, D. Loew, eds.) Friedr., Vieweg & Sohn, Braunschweig/Wiesbaden (1987) p 76.
34. A. H. Staib, D. Loew, S. Harder, E. H. Graul, R. Pfab, *Eur. J. Clin. Pharmacol* 30: 691 (1986).
35. A. H. Staib, B. G. Woodcock, D. Loew, O. Schuster, in *Novel Drug Delivery and its Therapeutic Application* (L. F. Prescott, W. S. Nimmo, eds.), John Wiley, Chichester (1989) p 89.
36. S. Harder, U. Fuhr, D. Beermann, A. H. Staib, *Br. J. Clin. Pharmacol.* 30: 35 (1990).
37. J. Hirtz, *Pharm. Int.* 5: 175 (1984).
38. A. Lambert, F. Vaxman, F. Crenner, T. Wittmann, J. F. Grenier, *Med. Biol. Eng. Comput.* 29: 191 (1991).
39. D. G. Colin-Jones, R. Cockel, K. F. R. Schiller, *Clin. Gastroenterol.* 7: 775 (1978).
40. K. E. Andresson, L. Nyberg, H. Dencker, J. Göthlin, *Eur. J. Clin. Pharmacol.* 9: 39 (1975).
41. H. Ochs, G. Bodem, P. K. Schäfer, G. Kodrat, H. J. Dengler, *Eur. J. Clin. Pharmacol.* 9: 95 (1975).
42. K. H. Antonin, P. Bieck, M. Scheurlen, M. Jedrychowski, H. Malchow, *Br. J. Clin. Pharmacol.* 19: 137S (1985).
43. C. H. Gleiter, K. H. Antonin, P. Bieck, J. Godbillon, W. Schönleber, H. Malchow, *Gastrointest. Endos.* 31: 71 (1985).
44. D. Brockmeier, H. G. Grigoleit, H. Leonhardt, *Eur. J. Clin. Pharmacol.* 29: 193 (1985).
45. D. Brockmir, H. G. Grigoleit, H. Leonhardt, *Eur. J. Clin. Pharmacol.* 30: 79 (1986).
46. H. Rameis, R. Pötzi, R. Kutscher, *Z. Gastroenterol.* 29: 1 (1991).
47. D. G. Maxton, P. J. Whorwell, *Aliment. Pharmacol. Ther.* 4: 305 (1990).
48. A. Stiehl, R. Raedsch, S. Walker, G. Rudolph, P. Klöters, in *Bile Acids and the Liver* (G. Paumgartner, A. Stiehl, W. Gerock, eds.), MPT, Lancaster (1987) p 323.
49. S. Walker, A. Stiehl, R. Raedsch, P. Klöters, *Gastroenterology* 90: 1779 (1986).
50. R. Balasubramanian, K. B. Klein, A. W. Pittman, S. H. T. Liao, J. W. A. Findlay, M. F. Frosolono, *J. Clin. Pharmacol.* 29: 444 (1989).
51. K. H. Antonin, P. Bieck, in *Drug Absorption at Different Regions of the Human Gastro-intestinal Tract: Methods of Investigation and Results* (N. Rietbrock, B. G. Woodcock, A. H. Staib, D. Loew, eds.), Friedr. Vieweg & Sohn, Braunschweig/Wiesbaden (1987) p 39.
52. P. C. Hawker, L. A. Turnberg, in *Large Intestine* (J. Alexander-Williams, H. J. Binder, eds.) Butterworths, London (1983) p 1.

53. C. H. Gleiter, C. Cremer, P. Bieck, N. Hengen, G. Kieninger, W. Schön-leber, *Z. Gastroenterol.* 22: 509 (1984).
54. S. A. Riley, I. A. Tavares, A. Bennett, V. Mani, *Br. J. Clin. Pharmacol.* 26: 173 (1988).
55. J. Højsted, K. Rubeck-Petersen, H. Rask, D. Bigler, C. Broen Christen-sen, *Eur. J. Clin. Pharmacol.* 39: 49 (1990).
56. K. D. Hooker, J. T. DiPiro, J. A. Stanfield, G. A. De Laurier, B. M. Lampert, J. T. Stewart, F. F. Knapp, *Clin. Pharm.* 8: 354 (1989).
57. A. G. de Boer, F. Moolenaar, L. G. J. De Leede, D. D. Breimer, *Clin. Pharmacokin.* 7: 285 (1982).

6

Comparison of In Vitro and In Vivo Dissolution for the Study of Colonic Drug Absorption

Dierk Brockmeier
Clinical Research, Hoechst AG, Frankfurt, Germany

I. INTRODUCTION

Drug therapy is predominated by oral formulations, since the oral route is the most convenient and popular means of applying a drug. The systemic availability of orally administered drugs is determined routinely as part of drug development during phase I, and later as part of in vivo quality control. The reduced systemic availability of some oral formulations may be explained, or even predicted, by their pharmacokinetic fate in the body after intravenous administration. However, the reduction in availability of many drugs or formulations is not fully understood. The reasons for this have not been further explored, but were often attributed to lacking absorption or incomplete dissolution.

Since oral formulations play such an important role in drug therapy, expressing availability as percentage alone is unsatisfactory. Interest should also focus on the reasons for the reduced availability. Particularly with sustained-release formulations or drugs with slow dissolution rates, it is important to know whether the drug is absorbed throughout the gastrointestinal tract at the same rate and extent or whether the rate and extent differs markedly from site to site. Since the residence time of a formulation within a certain part of the gastrointestinal tract is limited, a small or limited rate of absorption may affect the extent of absorption. In the same way, the rate of dissolution from a solid oral dosage form may change the extent of absorption, depending on the total residence time of the formulation in the gastrointestinal tract or on the residence time in the absorbing part.

Kübler (1) has transformed these considerations into a pharmacokinetic model and applied this model to findings with sulfhathiocarbamid, propicilline, xylose, and small amounts of ascorbic acid (1, 2). The complex "window" model of Kübler was later modified in several ways (3–5) and used to explain the concentration–time profiles of griseofulvin, buformin, and sulfisoxazole (4), piretanide, and furosemide (5, 6).

The authors developing various "window" models were not referring to certain segments of the gastrointestinal tract. However, the dominant site of absorption for some drugs was localized (2) and certain sites like the large bowel were shown to be nearly incapable of absorbing relevant amounts of drug (7, 8).

Experimental results with theoretical considerations on the physiology of the gastrointestinal tract led to the assumption that drug absorption occurs primarily in the upper part of the gastrointestinal tract. In the anatomical reserve length concept for intestinal absorption this presumption became part of the consideration, but is not necessarily absolutely linked to it (9).

Hirtz questioned the "window" concept because of the "absence of convincing experimental proof for its existence" (10) and his own results with metoprolol (11). Since then, equipment and experimental procedures have been developed or adjusted to allow for study of the absorption properties of different sites of the gastrointestinal tract generally, and of the colon particularly. They consist of intubation or endoscopy techniques, the high-frequency (HF) capsule, or the gamma-scintigraphic visualization technique.

Recently, a new approach was elaborated to study the absorption properties of the gastrointestinal tract, and particularly of the colon by comparing the in vitro results of drug dissolution with the in vivo profile of dissolution or absorption (12–15). It is a mathematically based approach that was deduced from moment analysis for both in vitro dissolution testing and in vivo studies of the same formulation. The results obtained by this method encouraged the hypothesis that controlled-release formulations can be used as probes to explore the rate and extent of absorption along the entire gastrointestinal tract.

Since the transit times of formulations through the different sections of the gastrointestinal tract have been studied extensively and are known, the result obtained by the method described here can also be used to specify which fraction of the dose is absorbed from the small intestine and which from the colon. Furthermore, it can be judged whether the rate constant of absorption is changing along the intestinal tract in general and with the passage of the formulation into the colon in particular. However, the controlled-release formulation can only function as a probe if a relevant amount of the dose is still delivered when it has reached the colon.

II. METHODOLOGIC BACKGROUND

The new approach focuses on continuous (point by point) comparison of the in vivo with an in vitro dissolution profile. However, it must be stressed that the in vivo dissolution profile can only be determined indirectly (e.g., by deconvolution). Therefore, it seems more appropriate to use the term *hypothetical in vivo dissolution profiles*.

Candidates for such a comparison are only formulations that show a sustained- or controlled-release profile in vitro that fulfill certain prerequisites (12). For example, the in vitro dissolution profile must characterize the formulation, but not the dissolution apparatus used. These prerequisites will be discussed later.

In general, the in vitro and in vivo dissolution profiles are not immediately superimposable. Comparison of in vitro and in vivo dissolution profiles starts with the assumption, which must be confirmed or rejected, that the in vivo dissolution curve has in principle the same profile as the in vitro curve. They differ only in the absolute units of their time axes, that is, both dissolution profiles can be superimposed by linear transformation of the time axes of either (13). This concept is illustrated in Figure 1.

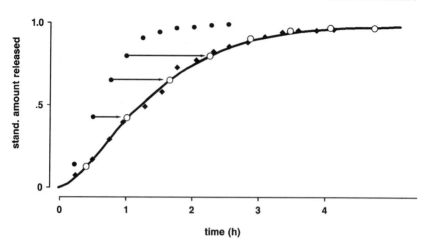

Figure 1 Homomorphism of two dissolution profiles and equivalence of the experimental conditions. Results shown are from dissolution testing of film-coated tablets of molsidomine in the Sartorius (●) and the USP paddle (◆) dissolution model. Least-square adjustment of a cumulative gamma function from data with the paddle model (solid line). Release profiles are superimposable by uniform linear transformation of the time base (arrows). The transformed data (○) are randomly distributed around the gamma function, showing that the two dissolution profiles provide the same information.

Figure 1 shows the cumulative amount released under two different in vitro dissolution conditions, A and B, and how both are correlated. The same considerations are valid if one of the two profiles represents a hypothetical in vivo dissolution. A cumulative gamma function, depicted as a solid line, has been adjusted to the data for condition B. After one rescales the time values of all observations related to condition A by uniform linear transformation, the data represented by open circles are obtained. This linear transformation is illustrated for three data pairs by arrows. The transformed data for condition A are randomly distributed around the function adjusted to the data for condition B.

The profiles are perfectly superimposable after linear transformation of the time bases of either. They are, therefore, homomorphic and the dissolution conditions are termed equivalent (13).

Dissolution profiles superimposable without transformation of the time axes are termed isomorphic (16).

This method of comparing two dissolution profiles is the essential feature of continuous in vitro–in vivo correlation and any conclusions based on it. It simply means that the elapse times under conditions A and B are different. There is an easy way to show homomorphism of two dissolution profiles without describing the mathematics. The procedure is illustrated in Figure 2.

The aim of the technique is to determine the time at which the same amounts have been dissolved under both conditions (14). For example, one starts with a certain amount dissolved under condition A at the t_A and determines the same amount dissolved under condition B at time t_B, if necessary by interpolation. Then follows a check of whether t_A and t_B are related linearly for the entire dissolution profiles by plotting t_B against t_A.

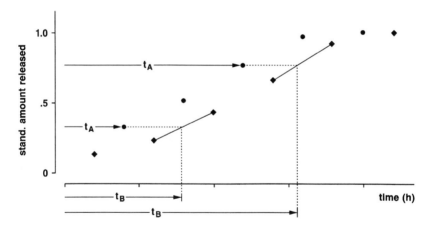

Figure 2   Technique for testing homomorphism of dissolution profiles. Cumulative fractions of drug dissolution as functions of time (●, ◆) originate from experiments under different dissolution conditions or from different apparatuses. Homomorphism of both dissolution profiles (i.e., equivalence of apparatuses or conditions) was tested by correlating dissolution times t_B with τ_A related to identical amounts dissolved under condition A (●) and B (◆), respectively. Related time pairs were determined by suitable interpolation of the dissolution data; linear interpolation is shown (adapted from ref. 14).

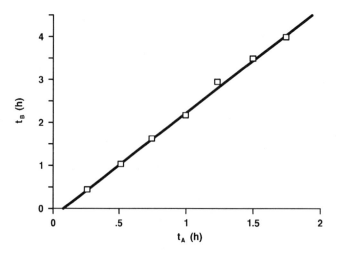

Figure 3 Correlation of dissolution times. Correlation of dissolution time, t_A and t_B, related to the same amounts dissolved under condition A and B for the data depicted in Figure 1. Intercept and slope provide the parameters to scale the time axis related to condition A to the time axis related to condition B (adapted from ref. 13).

This analysis was carried out for the two dissolution curves in Figure 1; the result is shown in Figure 3. Since both profiles are homomorphic, the time related to the same amounts dissolved under both conditions is clearly linearly related. As rule, if two dissolution profiles are homomorphic, the time at which the same amounts are dissolved is linearly related and vice versa.

This type of plot has been called the Levy-plot by our working group (14) since Gerhard Levy suggested this technique. He stated that

> Another approach to interpretation of the data is to plot the time required for the absorption of a given fraction of the dose versus the time required for the same fraction to dissolve in vitro. Such a plot yields a straight line nearly intercepting the origin, which suggests that the lag time referred to above is very short (17).

Levy compared in vivo absorption and in vitro release intuitively after the adjustment of in vitro dissolution to in vivo dissolution by varying the in vitro conditions. If the absorption of an aqueous solution is rapid, this correlation is correct.

Based on a more rigorous, mathematical definition of homomorphism of dissolution profiles, alternative algorithms to the Levy plot analysis can be used (13, 14).

III. APPLICATION

The two methods of continuous in vitro–in vivo correlation are illustrated in Figure 4.

A. First Alternative: From in Vitro Dissolution to in Vivo

Let us consider that a formulation is available and shows homomorphic dissolution profiles under various in vitro dissolution conditions. In general, each of these dissolution profiles has to be re-

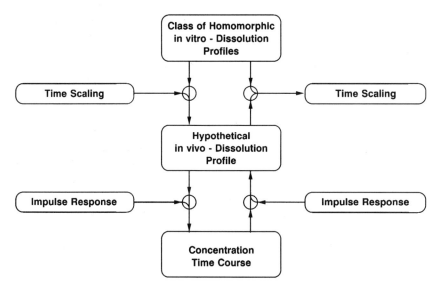

Figure 4 Two alternatives for the continuous comparison of in vitro and in vivo results. Left side of the scheme shows sequence of computational steps with the aim of comparing a predicted concentration profile with the actual readings. Right side shows procedure to obtain the hypothetical in vivo dissolution profile, which is then compared with the in vitro dissolution profile. See text for further details (adapted from ref. 15).

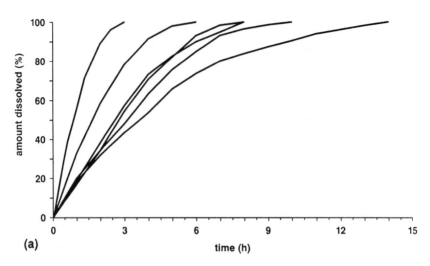

Figure 5 In vitro dissolution profiles for a sústained-release theophyl-line formulation. (a) Cumulative dissolution profiles of the sustained-re-lease theophylline formulation. Depending on the dissolution condition, duration of dissolution ranged from 3 to 14 hr. Dissolution was tested under various in vitro conditions including different apparatuses (Sarto-rius, paddle, rotating flask), different revolutions per minute or agitation modes, and different dissolution media. Profiles represent means of five samples originating from the same batch.

scaled individually to be superimposable with the hypothetical in vivo dissolution profile (first step on left of Fig. 4). The correct transformation of the time axis is essential. After transformation, either of these profiles can be used to predict the concentra-tion–time course by convolution. Convolution is a mathematical technique that allows one to predict the concentration–time profile (18, 19). For this it is necessary to know the input of drug to the body system (hypothetical in vivo dissolution, Fig. 4) and the response of the body system to an instantaneous input. In the context described here, the instantaneous input would be an oral solution and the concentration–time profile following its administration (standard-ized by dose) would be the response (impulse response, Fig. 4). Under certain conditions the concentration–time profile following intravenous application can substitute for that after an oral solu-tion to deduce the impulse response.

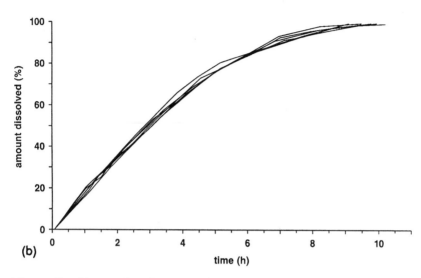

Figure 5   (b) Five dissolution profiles were rescaled to the time axis of the sixth (arbitrarily chosen as the reference) according to the methods for testing homomorphism of profiles (see ref. 16). Transformed profiles are practically superimposable (i.e., differences at each sampling point are less than the difference regularly observed between samples originating from the same batch) (adapted from ref. 22).

This first alternative is illustrated by data taken from a study with a sustained-release formulation of theophylline tested in eight healthy volunteers (20, 21). The sustained-release formulation and an oral solution was administered in a randomized crossover manner. The in vitro release of theophylline from this sustained-release formulation was tested under different conditions with different equipment (22). The resulting dissolution profiles are shown in Figure 5a. The release profiles are markedly different depending on buffer and agitation.

From this result it becomes apparent that one cannot specify which of the in vitro dissolution profiles, if any, is like the in vivo dissolution. However, for this formulation the dissolution profiles were homomorphic since they were adequately superimposable by linear transformation of the time axes (Fig. 5b). Since the in vitro dissolution profiles were homomorphic, it can be concluded that they reflect mainly a formulation property and not the dissolution condition. This is one of the more stringent requirements that must be fulfilled by any formulation used in the continuous in vitro–in vivo correlation. None of the untransformed in vitro dissolution profiles was like the in vivo dissolution profile. This can be easily shown by their convolution with the impulse response function, followed by comparison with the actual concentration–time data (15). The result of convolution is shown in Figure 6.

None of the predicted profiles fits the true observations. In each case the in vitro release is faster than the in vivo. The time axis of the in vitro dissolution profiles must therefore be transformed to become identical with the in vivo, that is, the in vitro clock and the in vivo do not run at the same speed. Transformation of the time axis of the in vitro dissolution profiles is necessary. Since all in vitro profiles are homomorphic, one can choose any of them as

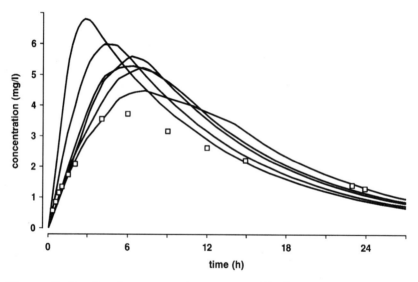

Figure 6 Predicted concentration–time profile for a sustained-release theophylline formulation without scaling the in vitro time towards the in vivo. Open squares (□) represent the mean serum concentrations of theophylline after administration of a sustained-release formulation. Solid lines are the results of prediction by convolution, with the various untransformed in vitro dissolution profiles as input functions. Body system was characterized by a weighting function, which is the dose-standardized and lag-time-corrected response to an oral solution (adapted from ref. 15).

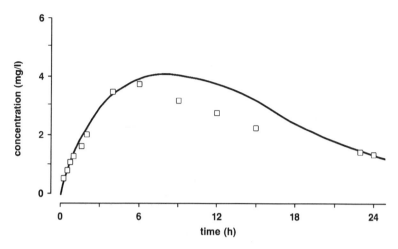

Figure 7 Predicted concentration–time profile for a sustained-release theophylline formulation with scaling of the in vitro time towards the in vivo. Open squares (□) represent the mean serum concentrations of theophylline. The solid line represents the result of prediction by convolution taking the time-scaled in vitro dissolution profiles as input function into the body system (adapted from ref. 15).

in vitro reference. The parameters for the transformation can be computed from the statistical moments of dissolution and transit times (15). After the time axis was rescaled, the prediction of the concentration–time course by convolution was reasonable, but not perfect (Fig. 7). Between 6 and 24 hr, the prediction differed clearly from the actual observations. This mode of analysis does not indicate any reason for the difference. Therefore, the alternative approach was applied to compare in vitro with in vivo results (right part of Fig. 4) (15).

B. Second Alternative: From in Vivo Concentration–Time Course to in Vitro

The hypothetical in vivo dissolution profile was estimated by deconvolution (first step on right of Fig. 4). The inversion of the convolution, that is, the determination of the input of drug to the body system, is termed deconvolution (19, 23–25). The result of deconvolution (the hypothetical in vivo dissolution profile) was then compared with one of the in vitro dissolution profiles by the Levy plot technique (second step on right of Fig. 4).

Levy plots were produced for each individual. They did not show a single linear relation between the time at which the same amount was dissolved under in vitro and in vivo conditions. Therefore, the in vitro and the hypothetical in vivo dissolution profiles are not homomorphic. However, the plots showed two sections, each of which could be described by a straight line. The same type of Levy plot was determined for each of the individuals. They differed in the slopes for the first and second section of the Levy plot and in the ratios between the two slopes (15).

The results for the two volunteers with the smallest and largest ratios between slopes are shown in Figure 8.

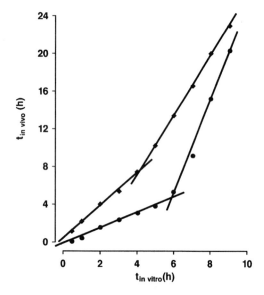

Figure 8 Correlation of in vitro and in vivo dissolution times for a sustained-release theophylline formulation. Correlation of dissolution times, $t_{in\ vitro}$ and $t_{in\ vivo}$, related to identical amounts dissolved under both conditions for the two subjects (◆ and ●), with smallest and largest ratio between the slopes for the two sections of the biphasic linear correlation. Same pattern for all subjects. The in vitro and the hypothetical in vivo dissolution profiles are not homomorphic and the dissolution conditions are apparently not equivalent (from ref. 15).

The transition from the first linear section to the second occurs, on average, after 5.3 hr (15), which is essential for the following. Possible theoretical reasons for such biphasic continuous in vitro–in vivo correlation include increasing clearance while absorption is still in progress, decreasing agitation in the gastrointestinal tract before complete in vivo dissolution, pH-dependent change in solubility, site-dependent degradation in the gastrointestinal tract, decreasing diffusion within the chymous mass, or decreasing absorption rate constant. Each of these has been discussed in detail (15) and most of them can be ruled out for theophylline. Therefore, it was concluded that most probably the rate constant of absorption for theophylline changes along the gastrointestinal tract and that this change occurs, on average, after 5.3 hr.

This interpretation was later confirmed by Staib and co-workers (26, 27). They used the HF capsule to release a theophylline solution at different sites of the gastrointestinal tract and determined the rate and extent of absorption. The extent of absorption (as measured by the area under the curve) was nearly the same for all sites of administration. However, the rate of absorption (as measured by the absorption half-life) differed to sixfold at the different sites (Table 1).

In the present example the time of transition from the first to the second linear section of the continuous in vitro–in vivo correlation apparently reflects the passage of the formulation from the small intestine to the ascending colon. The time of transition for this study was, on average, 5.3 hr. The average orocecal transit time reported for different formulations using the scintigraphic technique is about 5 hr (28–37).

IV. HYPOTHESES

The continuous in vitro–in vivo correlation was applied to data from a study with a sustained-release formulation of theophylline that had homomorphic in vitro dissolution profiles. In contrast to the expectations based on in vitro studies, the results showed that the hypothetical in vivo dissolution profile was not homomorphic with the in vitro dissolution profiles. The difference was explained by a possible change in the rate constant of absorption during transit of the formulation along the gastrointestinal tract (15). This

Table 1 Pharmacokinetic Characteristics of Theophylline

	N	AUC (mg h/l)	$t_{1/2.abs}$ (min)	$t_{1/2.elim}$ (hr)	MT (hr)
Oral solution	8	16.1 ± 3.6	9.5 ± 6.4	5.7 ± 1.4	8.5 ± 2.0
Stomach	3	16.3 ± 6.5	13.8 ± 8.8	6.3 ± 2.5	9.3 ± 3.6
Ileum	10	14.0 ± 4.5	25.6 ± 17.1	5.8 ± 1.3	9.0 ± 1.8
Ascending colon	6	15.0 ± 6.7	39.6 ± 27.6	7.4 ± 1.9	11.6 ± 3.1
Descending colon	4	14.4 ± 5.7	39.6 ± 36.4	6.9 ± 2.2	11.2 ± 3.2
Sigmoid colon	7	14.4 ± 4.8	64.0 ± 30.6	6.5 ± 1.7	10.8 ± 2.6

Source: Compiled from ref. 27.
Data were deduced from the concentration–time profile after administration of 100 mg theophylline at different sites of the gastrointestinal tract using the HF capsule and after administration of an oral solution.
AUC, Area under the concentration–time data curve; $t_{1/2.abs}$, absorption half-life; $t_{1/2.elim}$, elimination half-life; MT, total mean residence time.

interpretation was confirmed by Staib and co-workers using the HF capsule (26, 27). Therefore, the following hypotheses were formulated: Provided that a sustained-release formulation shows at least homomorphic in vitro dissolution profiles under different dissolution conditions. If a biphasic or polyphasic correlation is obtained when applying the continuous in vitro–in vivo correlation to this formulation, it suggests that the rate constant of absorption is different for the different parts of the gastrointestinal tract.

The larger the ratio of the slopes for the separate linear sections of the continuous in vitro–in vivo correlation, the larger the difference in the rate constant of absorption between the different sites of the gastrointestinal tract for the drug studied.

The region of the gastrointestinal tract in which the change of the rate constant of absorption takes place can be deducted from the continuous in vitro–in vivo correlation, that is, from the in vivo time related to the transition from the first linear section to the second.

V. SCRUTINY OF THE HYPOTHESES

A. Further Results for Theophylline

In a study with eight healthy volunteers, an oral solution of theophylline was compared with a theophylline formulation, which by content and release rate should have been suitable for once daily administration. The study followed a randomized crossover design. The in vitro dissolution of the formulation was determined by different apparatuses and under different conditions comparable to those described in Figure 5. Homomorphism was proven for the resulting dissolution profiles (38).

The concentration–time curve after administration of the 24 hr formulation was deconvoluted using the dose-standardized response after administration of the solution. The resulting hypothetical in vivo dissolution profile was subjected to Levy plot analysis; the in vitro dissolution profile obtained with a Sartorious apparatus was chosen as the in vitro reference. However, any of the in vitro dissolution profiles could have been used as reference since they were homomorphic with each alternative.

The correlation of time related to the same amounts dissolved under in vitro and in vivo conditions was biphasic, as already shown in Figure 8. A biphasic correlation was obtained for all subjects but one. The correlation for the mean curve is depicted in Figure 9, together with the mean for the study described first.

As can be seen in Figure 9, the slopes for the first and second section of the correlation plot differ for the two studies, although in both cases the same dissolution conditions from the Sartorious dissolution model were used. However, on average the time for the transition from the first to the second section is very similar for the two studies: 5.3 and 5.9 hr. Also, a large interindividual variation of the transition time was seen in both studies: 2.6–8.2 hr for the first study and 4.9–7.7 hr for the second study. The ratio of the slopes for the first and the second linear section of the correlation for both studies was broadly the same: 4.4 and 4.7 hr. In the first study 64% of the dose was dissolved and absorbed during the time period represented by the first section of the Levy plot, that is, 0–5.3 hr. Twenty-five percent of the dose was absorbed during the second section, which reflects absorption from the colon. Due to the slower release from the aimed 24 hr formulation, 41% of the

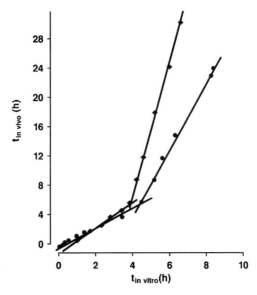

Figure 9 Correlation of in vitro and in vivo dissolution times for two different sustained-release theophylline formulations. Continuous in vitro–in vivo correlation for two studies with different sustained-release theophylline formulations. Mean values for both studies ($\blacklozenge$ and $\bullet$). For both formulations, correlation of dissolution times $t_{in\ vitro}$ and $t_{in\ vivo}$, related to identical amounts dissolved under in vitro and in vivo conditions showed the biphasic linear relation. Slopes differed slightly from one formulation to the other. Times for the transition from the first linear phase to the second were nearly identical.

dose was dissolved and absorbed during the first time period of the Levy plot (0–5.9 hr), and 20% during the colonic phase. The formulations tested in the two studies differed clearly in content and rate of release. However, the same pattern of continuous in vitro–in vivo correlation was obtained and the parameters describing these correlations were very similar.

These results show that reasonable amounts of theophylline have already been absorbed before the formulations reached the colon. However, the absorption of theophylline is not restricted to the upper intestine; it continues when the formulation reaches the colon. Therefore, the passage of a sustained-release formulation into the colon is not the sole cause for reduced bioavailability in the case of the 24 hr-formulation. For this formulation the release was too slow and probably the entire dose was not released during passage through the gastrointestinal tract.

B. Testing the Hypotheses for Piretanide

The best way to confirm our hypotheses (section IV) seemed to be the use of a drug for which a change in absorption along the gastrointestinal tract had been verified and for which also a sustained-release formulation had been studied under in vitro and in vivo conditions. Piretanide, a loop diuretic, is a drug that meets both requirements. Figure 10 shows the concentration–time curve of piretanide after endoscopic administration of a 3 mg solution into the stomach, duodenum, and ascending colon (8).

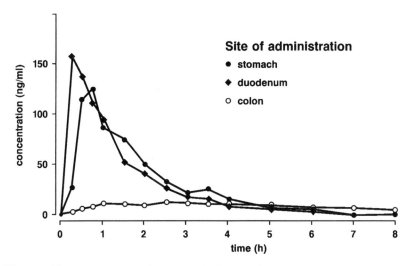

Figure 10 Absorption of piretanide from different sites of the gastrointestinal tract. Median serum concentration–time profiles of piretanide after endoscopic administration at three different sites of the gastrointestinal tract. About the same amount is absorbed when piretanide is instilled into the stomach and into the duodenum. A striking difference is observed (amount and rate) when piretanide is instilled into the ascending colon (from ref. 8).

The absorption of piretanide was delayed after gastric administration compared to intraduodenal. After administration of this agent into the duodenum, the concentration–time curve was practically like that after intravenous administration, indicating extremely rapid absorption. However, a remarkable difference in the concentration–time profile was observed when the drug was administered into the colon. Only 25% compared with intraduodenal instillation is very slowly absorbed. In one subject, the concentration never exceeded the detection limit. The concentration–time profiles for the remaining subjects resembled very much those seen after continuous infusion.

According to these absorption properties of the gastrointestinal tract, piretanide must show a biphasic Levy plot in the continuous in vitro–in vivo correlation to support our hypotheses. If not, the new approach would be invalid.

In a study with 27 volunteers, piretanide was administered intravenously and orally as conventional tablet and as an experimental sustained-release formulation following a randomized crossover design (39, 40). The dose for all routes was 6 mg. Data after administration of an oral solution were not available. Therefore, the hypothetical in vivo dissolution profile for the sustained-release formulation was estimated by deconvolution using the dose-standardized profile for intravenous application as impulse response. This is not a serious drawback. It only implies that the result of deconvolution cannot be regarded as the hypothetical in vivo dissolution profile, but encompasses the absorption profile. The in vivo dissolution profile obtained by deconvolution is always hypothetical and represents the actual in vivo dissolution profile only in cases in which the absorption rate constant does not change along the gastrointestinal tract. For deconvolution, the impulse-response function used is assumed to be valid for the entire gastrointestinal tract. Since the concentration–time response after administration into the duodenum is almost the same as that after intravenous bolus application, the use of the response after intravenous administration in the continuous in vitro–in vivo correlation equates to the situation in which the impulse response has been deduced from administration of a solution into the duodenum.

The hypothetical in vivo dissolution profile obtained differed clearly from the in vitro dissolution profile. Application of the continuous in vitro–in vivo correlation resulted in a biphasic Levy

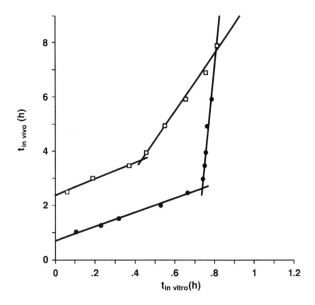

Figure 11 Correlation of in vitro and in vivo dissolution times for a sustained-release piretanide formulation. Correlation is biphasic, as expected from the absorption characteristics of the gastrointestinal tract for piretanide (see Fig. 10). Results for the two subjects with the smallest (□) and the largest (●) ratio between slopes for the two sections of the biphasic correlation are shown. Same pattern of biphasic linear correlation for all but one subjects. It is concluded that the rate constant of absorption changes abruptly after about 3 hr, with only small amounts of piretanide being absorbed thereafter.

plot. Figure 11 shows the two subjects with the largest and smallest ratio between the slopes for the first and second section of the correlation.

‧ This pattern was observed in 26 of the 27 subjects. The Levy plots differed only in the slopes for the first and second linear section and in the transition time, from the first to the second section. This time varied from 2.1 h to 5 hr, with an average of 3.6 hr (41).

These results clearly support the first hypotheses stated above. At least a biphasic Levy plot is obtained if the absorption rate constant changes along the gastrointestinal tract. Furthermore, the ratio between the first and second slope was 12 on average, in contrast to 4 for theophylline. The larger the difference in the rate constant of absorption for the different parts of the gastrointestinal tract becomes, the larger becomes the ratio between the slopes. Based on the information on gastrointestinal transit times compiled from other studies, it can be concluded that the time of change (3.6 hr) is too small to reflect the passage of the formulation from the small intestine to the colon.

Piretanide is absorbed mainly in the upper small intestine and only to a very small extent in the lower small intestine and colon. Therefore, its systemic availability is defined by the time the formulation resides in the well-absorbing part of the gastrointestinal tract (window). Figure 12 confirms this conclusion; it shows a posi-

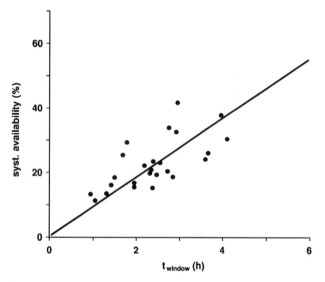

Figure 12 Correlation of the amount of piretanide absorbed with the time the formulation resides in the absorption window. Amount of piretanide absorbed from the lower part of the gastrointestinal tract is very small. Bioavailability of piretanide depends on the time the sustained-release formulation resides in the part with rapid and complete absorption (window). Bioavailability plotted over window time correlates well. The time the sustained-release formulation resides in the well-absorbing part of the gastrointestinal tract (window time) was read from the individual continuous in vitro–in vivo correlation.

tive correlation between systemic availability of piretanide and the time period in this window. The time for formulation resides in this window can be read from the continuous in vitro–in vivo correlation. However, the absorption capability is not entirely present or absent in the sense of an all-or-nothing characteristic, but the gastrointestinal tract shows a part with a rapid and almost complete absorption of piretanide and a part with very slow and, therefore, incomplete absorption.

The absorption of piretanide in the gastrointestinal tract can be assessed in a less invasive manner using the continuous in vitro–in vivo correlation. Furthermore, the average residence time of the formulation in the absorption window can be determined more precisely. Based on the knowledge of transit times of different formulations through the gastrointestinal tract, it can be concluded that piretanide is absorbed only to a small extent from the distal part of the small intestine and from the colon. This has a direct impact on the development of therapeutic formulations: the in vivo dissolution must be adjusted to the time the formulation resides in the well-absorbing upper part of the intestinal tract.

C. Isosorbide-5-Mononitrate

Another continuous in vitro–in vivo correlation was directly compared with the results obtained by gamma scintigraphy. Wildfeuer and co-workers have shown that the rate of absorption of isosorbide-5-mononitrate (ISN) from the colon was only slightly slower than from the upper intestine (42). The extent of absorption was, however, about the same for all sites of administration. They used the HF capsule that was opened in the stomach, duodenum, jejenum and ascending colon. Since the difference between the rate constants of absorption was small, ISN was a crucial candidate for testing the hypotheses stated, especially the sensitivity of the method. Applying continuous in vitro–in vivo correlation to ISN should therefore lead to a biphasic correlation, but only a small difference between the slopes of the two sections of the correlation is expected. Fischer and co-workers studied the absorption of ISN after administration of a slow-release pellet formulation (43). They characterized it as controlled-release because the in vitro release profile was independent of the dissolution conditions used (i.e., resulting in isomorphic dissolution profiles). They monitored the

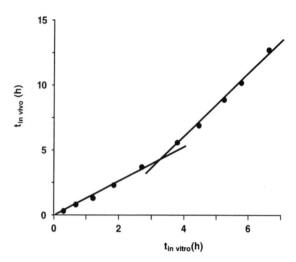

Figure 13 Correlation of in vitro and in vivo dissolution times for a controlled-release isosorbide-5-mononitrate (ISN) formulation. Correlation of dissolution times, $t_{in\ vitro}$ and $t_{in\ vivo}$, related to identical amounts dissolved under both conditions (mean values). According to expectation, a biphasic linear relation results for the controlled-release ISN formulation. The change of the rate constant of absorption for the upper and lower part of the gastrointestinal tract is less pronounced than with piretanide and theophylline. The ratio between the slopes for the two sections of the biphasic correlation is therefore smaller than for those substances.

intestinal transport of the ^{111}In-DTPA-labeled pellets by gamma scintigraphy after having proved that the release from labeled and unlabeled pellets was the same.

The results of the continuous in vitro–in vivo correlation were deduced from their figures on the in vitro and in vivo drug release and are shown in Figure 13.

Even the small difference between the rate constant of absorption from the upper intestine and from the colon was revealed by the continuous in vitro–in vivo correlation. The difference in the rate constant of absorption was not very pronounced. This also became obvious in the slopes for the first and second section of the Levy plot. They did not differ as clearly as for piretanide or theophylline. The ratio between both slopes was only 1.8. The time at which the first line passes to the second is 4.6 hr, which is about the same as the orocecal transit time. The data show that ISN is absorbed from the colon with only a small difference in the rate constant of absorption when compared to that for the upper intestine.

The arrival in the colon determined by Fischer et al. started at 4 hr. After 6.8 hr, an average of 50% of the pellets were found in the colon. They reported that the pellets accumulated at the ileocecal junction and entered the colon in one or a few boluses. The propulsion as bolus can explain the clear break in the transition from the first linear section in the Levy plot to the second.

VI. DISCUSSION

The theoretical and methodologic basis for the continuous in vitro–in vivo correlation (12–15) has been developed by elaborating the concept of in vitro–in vivo correlation using the mean in vitro dissolution time and the mean residence time in the body (44, 45). The results of a study with a sustained-release theophylline formulation have been analyzed according to this technique. In contrast to our expectation, this approach resulted in a biphasic linear in vitro–in vivo relationship (15). It was thought most probable that this is caused by a change in the absorption rate constant for theophylline along the gastrointestinal tract. This interpretation was later confirmed by Staib and co-workers (26, 27), who showed that the rate constant of absorption was slower, particularly when the drug was administered into the colon.

Studies on continuous in vitro–in vivo correlation call for sustained-release formulations with isomorphic or at least homomorphic dissolution profiles under different in vitro dissolution conditions. First, suitable formulations can be developed in vitro. In vivo studies designed to apply the methods of continuous in vitro–in vivo correlation would follow. Such a formulation must be studied in volunteers with an oral solution or an intravenous application. The findings of Staib et al. with theophylline encouraged the hypothesis that the continuous in vitro–in vivo correlation can be used to screen the absorption properties of the gastrointestinal tract for site-dependent (shown as time-dependent) differences in rate and amount. This hypothesis was confirmed for different sustained- and controlled-release formulations of theophylline (40, 41).

The hypothesis was further scrutinized by inverting it. Continuous in vitro–in vivo correlation should lead to at least biphasic, linear correlations for drugs with known changes of the absorption rate constant along the gastrointestinal tract. Two drugs were selected: piretanide, for which it has been shown that the rate constant of absorption changes extremely, and ISN, for which the change is only small. Our hypothesis was confirmed (40, 41) for both drugs. Further studies with sustained-release piretanide formulations showed the same biphasic in vitro–in vivo correlation, with remarkable consistency of the transition time from the first to the second section of the biphasic linear relation.

The time of transition from the first to the second section reflects generally the passage of the formulation from the site with a higher rate constant of absorption to the site with a lower rate constant. Knowledge of the gastrointestinal transit time of various formulations allows us to predict the anatomical site where the change in rate constant of absorption takes place.

Therefore, it is usually possible to judge whether the drug is absorbed from the large bowel or not. It requires a formulation that shows a sustained in vivo release exceeding the orocecal transit time. If the result shows that the drug is not or only barely absorbed from the colon, the release of drug has to be adjusted. The same is valid for drugs that are minimally soluble in the gastrointestinal tract. The continuous in vitro–in vivo correlation would indicate whether the residence time in the well-absorbing part of the gastrointestinal tract is sufficient to dissolve the dose administered.

With the method of continuous in vitro–in vivo correlation, patterns of biphasic linear correlation have been found for other drugs not reported here. Most of them showed a transition time that correlated with the orocecal transit time reported in the literature. The results also showed that in most cases the colon is capable of absorbing the drug well. In some cases the continuous in vitro–in vivo correlation resulted in Levy plots with three linear sections. Whether this indicates that the gastrointestinal tract consist of three parts with different rate constant of absorption must be confirmed by direct measurement using the intubation technique or the HF capsule.

In summary, it has been shown that the absorption features of the gastrointestinal tract, particularly of the colon, can be studied by using sustained-release formulations as pharmaceutical probes with application of the continuous in vitro–in vivo correlation. A two-step strategy should be adopted in the future. First, a formulation is developed to serve as pharmaceutical probe only. Second, the oral dosage form of that drug is optimized with consideration given to both the therapeutic goal and the site-dependent rate of drug absorption.

REFERENCES

1. W. Kübler, *Gastroenterologia* 104 (Suppl): 231–235 (1965).
2. W. Kübler, *Z. Kinderheilkd.*, 108: 187–196 (1970).
3. J. G. Wagner, in *Fundamentals of Clinical Pharmacokinetics*, Drug Intelligence Publications, Hamilton (1979) p 182.
4. R. Süverkrüp, *J. Pharm. Sci.* 68: 1395 (1979).
5. D. Brockmeier, H.-G. Grigoleit, H. Heptner, B. H. Meyer, *Methods Find Exp. Clin. Pharmacol.* 8: 731 (1986).
6. D. Brockmeier, *Methods Find Exp. Clin. Pharmacol.* 8: 593 (1986).
7. D. Loew, A. H. Staib, S. Harder, O. Schuster, E. H. Graul, in *Drug Absorption at Different Regions of the Human Gastrointestinal Tract: Methods of Investigation and Results* (N. Rietbrock, B. G. Woodcock, A. H. Staib, and D. Loew, eds.), Friedr. Vieweg & Sohn, Braunschweig/ Wiesbaden (1987) p 69.
8. D. Brockmeier, H.-G. Grigoleit, H. Leonhardt, *Eur. J. Clin. Pharmacol.* 30: 79 (1986).
9. N. F. H. Ho, H. P. Merkle, W. I. Higuchi, *Drug. Dev. Ind. Pharm.* 9: 1111 (1983).
10. J. Hirtz, *Br. J. Clin. Pharmacol.* 19: 77 S (1985).
11. J. Godbillon, D. Evard, N. Vidon, M. Duval, J. P. Schoeller, J. J. Bernier, J. Hirtz, *Br. J. Clin. Pharmacol.* 19: 113 S (1985).
12. D. Brockmeier, *Acta Pharm. Technol.* 32: 164 (1986).
13. D. Brockmeier, H. M. von Hattingberg, *Arzneimittelforsch./Drug Res.* 32: 248 (1982).
14. D. Brockmeier, D. Voegele, H. M. von Hattingberg, *Arzneimittelforsch./Drug Res.* 33: 598 (1983).
15. D. Brockmeier, H. J. Dengler, D. Voegele, *Eur. J. Clin. Pharmacol.* 28: 291 (1985).
16. W. R. Ashby, in *An Introduction to Cybernetics*, John Wiley & Sons, New York (1956), p 102.
17. G. Levy, *Arch. Int. Pharmacodyn.* 152: 59 (1964).
18. A. Rescigno, G. Segre, in *Drug and Tracer Kinetics*, Blaisdell Waltham/ Toronto/London (1966) p 102.

19. E. Stepanek, in *Praktische Analyse linearer Systeme durch Faltungsoperationen*, Akad. Verlagsges. Geest & Porting KG, Leipzig (1976).
20. H. J. Dengler, I. Beuckelmann, A. Türk, D. Voegele, in *Theophylline and Other Methylxanthines* (N. Rietbrock, B. G. Woodcock, and A. H. Staib, eds.), Friedr. Vieweg & Sohn, Braunschweig/Wiesbaden (1982) p 83.
21. I. Beuckelmann, Die Verwendung der mittleren Verweildauer zur Beurteilung der absoluten und "relativen" Bioverfügbarkeit eines Theophyllin Retard Präparates, Doctoral thesis, University of Bonn (1983).
22. D. Voegele, Zur Bedeutung der Verteilungsfunktion der Freisetzungszeiten für die Galenik, Doctoral thesis, University of Düsseldorf (1985).
23. F. Langenbucher, *Pharm. Ind.* 44: 1275 (1982).
24. F. Langenbucher, *Pharm. Ind.* 44: 1166 (1982).
25. G. T. Tucker, *Acta Pharm. Technol.* 29: 159 (1983).
26. A. H. Staib, D. Loew, S. Harder, E. H. Graul, R. Pfab, *Eur. J. Clin. Pharmacol.* 30: 691 (1986).
27. A. H. Staib, D. Loew, S. Harder, J. Kollath, E. H. Graul, O. Schuster, B. Hugemann, in *Drug Absorption at Different Regions of the Human Gastrointestinal Tract: Methods of Investigation and Results* (N. Rietbrock, B. G. Woodcock, A. H. Staib, D. Loew, eds.) Friedr. Vieweg & Sohn, Braunschweig/Wiesbaden (1987) p 53.
28. F. N. Christensen, S. S. Davis, J. G. Hardy, M. J. Taylor, D. R. Whalley, C. G. Wilson, *J. Pharm. Pharmacol.* 37:91 (1985).
29. S. S. Davis, J. G. Hardy, M. J. Taylor, D. R. Whalley, C. G. Wilson, *Int. J. Pharm.* 21: 167 (1984).
30. S. S. Davis, J. G. Hardy, M. J. Taylor, D. R. Whalley, C. G. Wilson, *Int. J. Pharm.* 21: 331 (1984).
31. J. G. Hardy, C. G. Wilson, E. Wood, *J. Pharm. Pharmacol.* 37: 874 (1985).
32. J. C. Maublant, M. Sournac, J.-M. Aiache, C. G. Veyre, *J. Nucl. Med.* 28: 1199 (1987).
33. G. Parker, C. G. Wilson, J. G. Hardy, *J. Pharm. Pharmacol.* 40: 376 (1988).
34. A. F. Parr, R. M. Beihn, R. M. Franz, G. J. Szpunar, M. Jay, *Pharm. Res.* 4: 486 (1987).
35. M. Sournac, J. C. Maublant, J.-M. Aiache, C. G. Veyre, J. Bougaret, *J. Control Release* 7: 139 (1988).
36. C. G. Wilson, G. D. Parr, J. W. Kennerley, M. J. Taylor, S. S. Davis, J. G. Hardy, J. A. Rees, *Int. J. Pharm.* 18: 1 (1984).
37. C. G. Wilson, J. G. Hardy, *J. Pharm. Pharmacol.* 37: 573 (1985).
38. D. Voegele, D. Brockmeier, H. M. von Hattingberg, Modelling of input function to drug absorption by moments, in *Proceedings of the Sympo-*

sium on Compartmental and Noncompartmental Modeling in Pharma-cokinetics, Smolenice Castle, CS (1988) p 29.

39. Dagrosa, E. , Hajdú, P., Malerczyk, V., Rupp, W., Damm, D. in, *Pharmacokinetics of Pretanide (HOE 118) after Single Dose of 6 mg Orally and Intravenously to Healthy Men*, Internal Report Hoechst AG, Frankfurt (1981).

40. D. Brockmeier, *Eur. J. Clin. Pharmacol.* 36 (Suppl.): A34 (1989).

41. D. Brockmeier, Analyse der Zeitabhängigkeit der Resorption von Arzneistoffen aus dem Gastrointestinaltrakt durch in vitro–in vivo Korrelation der Arzneistofffreisetzung, Habilitation thesis, University of Giessen (1990).

42. A. Wildfeuer, H. Laufen, R. Dölling, G. Pfaff, B. Hugemann, H. E. Knoell, O. Schuster, *Therapiewoche* 36: 2996 (1986).

43. W. Fischer, A. Boertz, S. S. Davis, R. Khosla, W. Cawello, K. Sandrock, G. Cordes, *Pharm. Res.* 4: 480 (1987).

44. H. M. von Hattingberg, D. Brockmeier, D. Voegele, in *Methods in Clinical Pharmacology* (N. Rietbrock, G. Woodcock, G. Neuhaus, eds.), Friedr. Vieweg & Sohn, Braunschweig/Wiesbaden (1979) p 85.

45. D. Voegele, H. M. von Hattingberg, D. Brockmeier, *Acta Pharm. Technol.* 27: 115 (1981).

7

Osmotic Systems for Colon-Targeted Drug Delivery

**Felix Theeuwes, Patrick L. Wong,
Terry L. Burkoth, and Deborah A. Fox**
Alza Corporation, Palo Alto, California

I. INTRODUCTION

In this chapter we will review dosage forms for drug delivery to the colon, with emphasis on osmotic delivery systems, and describe some desirable features of such systems. We will consider agents that could provide better therapy with colonic delivery or agents with an improved profile of colonic delivery. Key physiological variables related to colon–targeted dosage form design will be examined. Our discussion will include detail on state-of-the-art osmotic systems with general applicability to colon targeting.

Two important motives make the colon a rational site for targeted drug delivery. The first is the need to treat serious localized chronic, often progressive and debilitating, disease states (1). These are the inflammatory bowel diseases, ulcerative colitis, and Crohn's disease. Their detailed causes are complex and largely unknown (2–4). Promising pharmacologic agents often lack specificity. For example, steroids also affect the function of the hypothalamic–pituitary–adrenal axis. This influences growth in children treated for Crohn's disease (5). Other disorders with major manifestations in the colon include the complex motility disorders of irritable bowel syndrome (6), diarrhea or constipation, and colon cancer. Transporting drugs directly to the colon to treat such conditions is highly desirable to focus therapy on the site of disease and to avoid unwanted systemic side effects. The second motivation for targeting drugs to the colon is to use this organ as an entry site for orally administered systemic drugs sensitive to conditions in the upper gastrointestinal tract, such as the acidic gastric environment or enzymatic activities in the small intestine. Proteins or peptides are the most notable drugs that fall into this class (see Chapter 8). The colon presents a less hostile environment to proteins than the upper gastrointestinal tract because it owns importantly less proteolytic activity. Gibson and co-workers (7) have shown that protein hydrolysis by intestinal contents from ileostomy subjects was 20–60 times greater than hydrolysis by fecal slurries from normal individuals (Fig. 1). Proteolysis in the colon was also qualitatively different from that in the small intestine. Evidence for this includes the range of proteins hydrolyzed and the susceptibility of the enzymes to protease inhibitors. Other investigations analyzing proteolytic activity in intestinal contents from victims of sudden death show

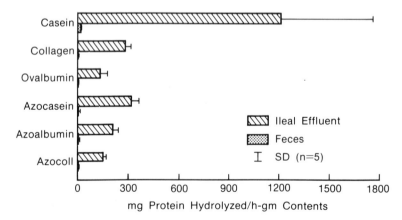

Figure 1 Hydrolysis of protein substrates by human ileal effluent and feces (adapted from ref. 7).

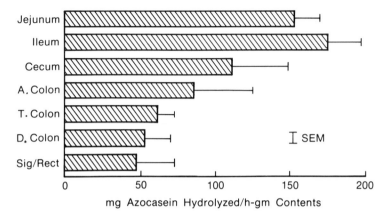

Figure 2 Protease activity in different regions of the human gastrointestinal tract. A. Colon, ascending colon; T. Colon, transverse colon; D. Colon, descending colon; Sig/rect, sigmoid colon/rectum (adapted from ref. 8).

that proteolytic activity decreases with progression down the gastrointestinal tract (8) (Fig. 2). These data, coupled with the fact that the distal colon has shown some permeability to proteins (9–12), suggest that targeting such agents to lower regions of the colon may reduce proteolytic degradation and enhance absorption.

The oral route is by far the most acceptable means of drug administration and offers the greatest promise for drug delivery in the colon, but also the greatest challenge. In the past, drug delivery has been specific to the lower bowel only when done via enemas and suppositories. These venerable dosage forms, aside from their stringent temporal and convenience limitations, cannot reach colonic regions proximal to the splenic flexure (13). Therefore, lesions in the proximal colon caused by Crohn's disease and ulcerative colitis cannot be effectively treated. The disadvantages of these methods include variability in the duration of drug delivery and of drug contact with colonic lesions or with the mucosa to achieve systemic absorption.

Eventually, enemas and suppositories may be relegated to use in testing a drug's therapeutic promise before a superior, orally administered, colon-targeted dosage form is designed. Such a dosage form must overcome the problems of gaining adequate access to the treatment site, promote patient compliance, ensure drug stability, and be commercially viable. To provide a general solution for local and systemic use, a colon-targeted system should be able to control the initiation of and duration of delivery throughout the colon.

II. ALIMENTARY TRACT TRANSIT TIMES

A review of transit and residence times in various sections of the gastrointestinal tract is essential for any discussion of key features of colon-targeted dosage form design. This discussion of how transit behavior may affect colon-targeted drug delivery is done in the context of efforts to treat inflammatory bowel diseases. Because these diseases can affect the entire colon, effective treatment calls for a dosage form that can be triggered to start drug delivery when it reaches the ileocecal junction. It should continue delivering throughout its duration of residence in the colon. Crohn's disease often gives rise to strictures of the intestine. Therefore, it appears wise to devise dosage forms comprising multiple small units to deliver the needed amount of drug, instead of a single unit of a large size. Therefore, any discussion of passage of a dosage form should take such choices into account.

Total transit time through the gastrointestinal tract from mouth to anus can be classified into three residence times associated with three distinct parts of the tract: the stomach, the small intestine, and the colon.

A. Gastric Residence Time

How long a solid dosage form can reside in the stomach is highly variable and dependent on several factors. Most notable are the object's size, possibly its density, and the subject's caloric intake. Given densities of 1–2 g/cm^3 for all drugs, and a requirement that dosage forms weigh less than 1 g for comfortable swallowing, caloric intake appears to be the most important factor. In both animals and humans there are two distinct modes of gastrointestinal motility: fasted and fed. In the fasted mode, the motility pattern, commonly called the migrating motor complex (MMC), has alternating cycles of activity and quiescence that traverse the gastrointestinal tract (14–15). The intense contractions of the MMC are called the burst, phase 3, or housekeeper waves. Following the ingestion of a meal, the fasted pattern is replaced by irregular contractions of intermediate amplitude, characteristic of the fed mode. Transit time of dosage forms through the gastrointestinal tract is therefore somewhat dependent on whether they are taken with food.

Early studies to determine the effect of particle size on gastric residence time were done in dogs by Hinder and Kelley (16). The results suggest that particles larger than 2 mm in diameter were not emptied from the fed stomach but were retained and emptied with the phase 3 contractions of the MMC. Later studies in dogs (17) and humans (18–20) have shown that much larger tablets (at least 7 mm) can be emptied from the fed stomach.

Caloric intake appears to be more important than size in determining gastric residence time. In a study by Khosla and co-workers (19), nondisintegrating tablets were prepared from ethylcellulose and a mixture with Amberlite ion-exchange resin labeled with technetium^{99m}. The tablets were compressed to have diameters and weights of 3.1 mm (20 mg), 4.0 mm (35 mg), 5.0 mm (55 mg), 6.2 mm (75 mg), and 7.1 mm (81 mg). Normal subjects were each administered 10 tablets of a single size immediately after a breakfast that satisfied three test conditions and caloric contents: light (1,500 kJ), medium (2,300 kJ), or heavy (3,500 kJ). Residence times in various sections of the gastrointestinal tract were evaluated by gamma scintigraphy. The disappearance of tablets from the stomach (as measured by percentage of radioactivity) was evaluated as a function of time, and by tablet arrival at the ileocecal junction, showing entry into the colon.

The 10 tablets appeared to exit the stomach in random fashion at rates that were independent of size but strongly dependent upon the caloric content of the meal. Tablets appeared to cluster at the ileocecal junction, with later spreading throughout the colon (Fig. 3). When data for all tablet sizes were combined and organized by meal size, Figure 4 was obtained. These data show that administering a dosage form containing multiple units with various-sized meals can significantly affect the units' distribution over time throughout the gastrointestinal tract. Also, Figure 4 shows that administering multiple units with a heavy meal allows one to time

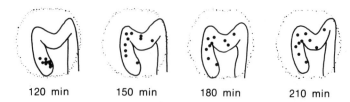

| 120 min | 150 min | 180 min | 210 min |

Figure 3 Distribution of 4 mm tablets in the colon 120, 150, 180, and 210 min after administration. The subject ate a light breakfast before taking the tablets.

their arrival. There is an interval of about 7.5 hr between arrival of the first and last units in the colon. With a light meal, arrival times would be distributed over only about 4 hr.

B. Small Intestinal Transit Time

Residence time in the small intestine is the period between the exit of foodstuffs or dosage forms from the stomach and their entry into the colon. Upon leaving the stomach, dosage forms pass from an environment of low pH ($\sim$1.2) into essentially neutral pH ($\sim$7.5).

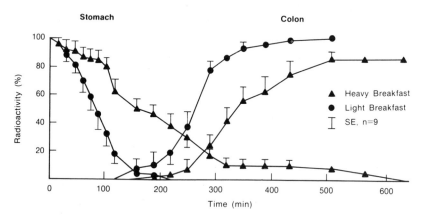

Figure 4 Effect of meal size on the transit of small tablets from the stomach to the colon.

Also, the small intestine is a highly absorptive area, where both passive and active transport mechanisms operate. Davis and co-workers (21, 22) used scintigraphic methods to measure the small intestinal transit time of solutions, pellets, and nondisintegrating tablets. The transit time in humans was remarkably independent of the size of the material (whether solutions, beads, or tablets) and of feeding conditions at ingestion. Intestinal transit time of pellets and tablet formulations (21) proved to be 3–4 hr, equal to that found for solutions containing ion-exchange beads (22). Analysis of the data in Figure 4, with calculation of the difference between arrival at the cecum and exit from the stomach, leads to a similar conclusion.

C. Colonic Transit Time

Transit or residence time of objects in the colon can be considered as the difference between mouth-to-anus transit time and mouth-to-cecum transit time. If one defines average mouth-to-cecum transit time as the time for 50% of ingested objects to reach the cecum, it appears that with a light breakfast or heavy breakfast, the average mouth-to-cecum transit time is about 3.5 or 6.5 hr (Fig. 4).

Mouth-to-anus transit times, among and within subjects, vary significantly. Figure 5 shows total transit times for nearly 1400 nondisintegrating OROS dosage forms in humans. Apparently, all systems were retained in the body for at least 10 hr. Ninety percent were retained for 24 hr, 50% for 30 hr, and less than 10% for longer than 75 hr. This shows that it would be futile to design oral delivery systems to deliver drug for much longer than 24 hr because a large portion of the systems might be excreted before the entire dose could be delivered.

D. Gastrointestinal Transit of Small Tablets in Patients with Ulcerative Colitis

The design of a drug delivery system to the large intestine for patients with active inflammatory bowel disease requires that one know system residence time in the gastrointestinal tract of such patients. Transit time through the small intestine and residence time in the colon are of specific interest. Hardy and co-workers

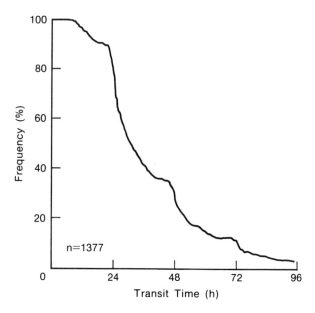

Figure 5   Survival frequency of OROS dosage forms in humans.

(23) studied six patients. Two had active and four had quiescent inflammatory bowel disease. All were receiving drug therapy. Non-disintegrating tablets of 4 mm diameter and 4 mm thickness were prepared from ethylcellulose coated with a cellulose acetate membrane. Each tablet contained a cation exchange resin radiolabeled with indium[111]. After an overnight fast, each subject ate a light breakfast and swallowed 5 tablets with 100 ml of water. Progression of tablets through the intestinal tract was evaluated by scintigraphy to measure gastric residence time, small intestinal transit time, and residence time in the ascending and transverse colon.

In general, the mean gastric residence time was about 1.6 hr. This is not different from that found in normal subjects. Transit time through the small intestine averaged 3.4 hr for patients with either active or quiescent disease. Again, this is within the range of normal subjects. Combined residence times in the ascending and transverse colon were 7.2 and 7.9 hr for the two patients with active disease, and 17–20 hr for patients with quiescent disease. Yet, even in patients with active disease, tablets were retained in the proximal colon for more than 6 hr and in the entire large bowel for at least 18 hr. It appears that total tablet transit was not abnormally fast in this group of patients with ulcerative colitis. Colonic residence time is surely prolonged enough to justify a system capable of continuous drug delivery.

In a later unpublished study, the same authors examined 11 more patients with active ulcerative colitis receiving drug therapy. They received the following regimen: breakfast at 8 a.m. and administration of a capsule containing five 5 mm nondisintegrating tablets; coffee and biscuits at 11:30 a.m.; lunch at 12:30 p.m.; coffee at 4 p.m.; and dinner at 7 p.m. The average gastric residence time was 1.3 hr, small intestinal transit time was 2.2 hr, and total transit time in all patients exceeded 22.5 hr. These data confirm the finding that total residence time in patients with ulcerative colitis does not differ significantly from that of healthy subjects.

From this review, it is clear that stomach residence time depends on caloric intake and is highly variable. Small intestine transit time, however, is relatively constant at 2–4 hr. Colonic residence time is long and variable, typically 10–30 hr.

III. COLON-TARGETED IMMEDIATE-RELEASE DOSAGE FORMS

Attempts have been made to achieve colon-targeted delivery from oral dosage forms via delayed, sustained, or controlled-release mechanisms. Enteric coatings have been used to block delivery in the stomach. Dissolution-controlled mechanisms have aimed to delay release of a drug to the colon and to ensure its delivery throughout the colon. These mechanisms, however, are known to be highly susceptible to pH, motility, and enzymatic activity, making accurate timing and delivery undependable (24).

Another approach has been the use of drugs chemically modified to make them latent in the upper alimentary tract, but released or activated enzymatically by colonic microflora (25, 26). Cross-linked polymers have likewise been developed to be degraded by the same enzymes or bacteria. When these are used as coatings on oral dosage forms, delivery of drug only in the colon results. Some examples of these approaches to the design of colon-targeted dosage forms are described here.

A. Prodrug-Based

The 60-year-old drug sulfasalazine is the powerful historic prece-
dent for the prodrug approach. Based on the colon's unique bacterial metabolism, it is designed to deliver drugs there selectively.
When sulfasalazine reaches the colon after oral administration, the
diazoreductases of the colonic bacteria cleave the azo bond, releasing two chemical entities, 5-aminosalicylic acid (5-ASA) and sulfapyridine, into the colonic lumen. It is thought that 5-ASA provides
local therapy to the colonic lesions of inflammatory bowel disease
while sulfapyridine is absorbed and is responsible for undesirable
side effects.

Because of these side effects, new 5-ASA prodrugs without the
pyridine sulfonamide congener have been devised. In reviewing
these newer 5-ASA-based drugs, Jarnerot (25) judged that olsalaz-
ine (Dipentum), bis-diazo-5-ASA, delivers 5-ASA most reliably to
the colon but is unsuitable for treating small bowel disease. Balsalazide is another prodrug undergoing phase III clinical trials that
is reputed to benefit from a less toxic carrier molecule than that in
sulfasalazine. Balsalazide is 5-ASA linked via a diazo moiety with
4-aminobenzoyl-β-alanine (13, 27).

Other reported prodrug approaches include the macromolecular
dextran esters of naproxen (28), which require two distinct enzymatic steps to liberate the activity entity, and the steroid glycosides
(29). These drugs rely on colon microflora to release active agly-
cones from their latent derivatives. It is not known if any of these
prodrugs have therapeutic value in patients.

B. Based on Degradation of Impermeable Membranes by Colonic Bacteria

Drugs can be delivered locally and selectively to the colon if they
are enclosed in a dosage form, such as a capsule, coated with an
azoaromatic cross-linked polymer subject to cleavage by azoreductases of the colonic microflora (9, 10). An example of such a protec-
tive coating is a random copolymer of styrene and hydroxyethyl
methacrylate, cross-linked with divinylazobenzene synthesized by
a free radical polymerization. Such coatings may be deposited by
solvent and are impermeable to the drug. The drug is protected
until the dosage form arrives in the large intestine, where the azo

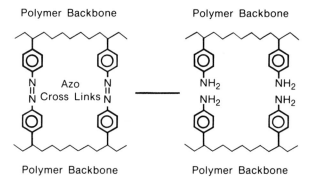

Figure 6 Reduction of azo polymer by colon bacteria. Only bacteria found in the colon can break the azo bond, disintegrating the film on the dosage form in the colon.

cross-links are broken enzymatically (Fig. 6). The coatings then disintegrate, exposing the soluble drug capsule for dispersion in the colon. This approach has been used to deliver insulin or vasopressin orally to rats using capsules or pellets coated with an azoaromatic polymer (9) and insulin to dogs using polymer-coated capsules (30).

C. Based on Degradation of Impermeable Membranes with Dual Protective Layers

A variation in dosage form design that also uses bacterial action in the colon to trigger drug release is shown in Figure 7. An outer enteric coating (i.e., cellulose acetate phthalate) protects the unit in the low pH in the stomach. When it enters the small intestine (pH 7.5), the enteric coating dissolves, exposing a polymeric coating: typically ethyl cellulose with microcrystalline cellulose and plasticizers. This polymeric coating is designed to stay intact in the small intestine until the dosage form enters the colon. There, bacteria are expected to digest the microcrystalline cellulose to allow for disintegration of the polymeric coating around the drug. X-ray photographs in a human study, using radiopaque barium sulfate placebos, confirmed that all the compressed tablets with the dual coating remained intact in the small intestine. Most of them (85%) disintegrated in the colon (31).

Polymeric coating
with microcrystalline
cellulose

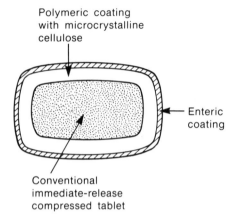

Enteric
coating

Conventional
immediate-release
compressed tablet

Figure 7   Cross-section of a multilayer, enteric-coated, immediate-release dosage form.

Compositions of 5-ASA are available commercially that use enteric coatings (26, 32) of Eudragit pH-sensitive acrylic resins. Asacol contains 5-ASA tablets coated with Eudragit-S. This breaks down at a pH above 7.0. Release may begin as early as in the midjejunum or in the terminal ileum. Salofalk and Claversol are 5-ASA tablets first coated with a semipermeable membrane of ethylcellulose and then with Eudragit-L (resin dissolves at a pH above 5.6). It is thought that 5-ASA delivery is inhibited in the stomach, duodenum, and proximal jejunum until the ethylcellulose coat is broken down. Then 5-ASA is released in the distal small intestine and in the colon. Pentasa is made of 5-ASA in microgranules coated with a semipermeable ethylcellulose membrane. There is little pH-dependent effect and drug release begins in the duodenum.

IV. COLON-TARGETED OSMOTIC THERAPEUTIC SYSTEMS

To deliver medication at a desired rate throughout the lumen or to a specific region of the colon, a well-defined controlled delivery system should function independent of the two gastrointestinal tract variables: gastric residence time and metabolism by bacterial flora. The system should not only avoid delivery in the upper gastrointestinal tract but should also deliver throughout the colon in a programmed manner. A family of osmotically driven systems is under development that satisfies these requirements.

To target drug delivery to begin at the ileocecal junction, we have used an enteric coating. This is impermeable to drug and blocks water transport into the osmotic system in the stomach, but it dissolves upon exposure to the higher pH of the intestine and starts a timing mechanism. This mechanism can be programmed to delay the start of osmotic drug delivery for periods up to 10 hours. For ulcerative colitis, drug delivery is timed to begin when the systems reach the cecum. Therefore, a 3–4 hr delay in startup of the osmotic dosage form is programmed to coincide with transit time through the small intestine. Upon arrival in the colon, the osmotic dosage form, divested of its enteric coating and hydrostatically primed by osmotic imbibition of water, starts drug delivery. Delivery can be programmed for any duration, depending on the indication. Delivery duration of 20 hr is required to ensure drug release throughout the distal sigmoid colon, where ulcerative colitis is most prevalent.

A. Drug-Dedicated Colon-Targeted Osmotic Therapeutic System

The OROS-CT osmotic therapeutic system can be used as a once–or twice–a–day colon-targeted dosage form. It can be used to treat diseases of the colon locally, or to direct drugs to the colon to achieve systemic absorption that is otherwise unattainable (33). The OROS-CT system can be a single osmotic unit or can comprise as many as 5 to 6 push-pull (34) units, each 4 mm in diameter, contained within a hard gelatin capsule (Fig. 8). Each bilayer push-pull unit contains an osmotic push layer and a drug layer, both surrounded by a semipermeable membrane. An orifice is drilled through the membrane next to the drug layer. Immediately after the OROS-CT is swallowed, the gelatin capsule containing the push-pull units dissolves. Because of its drug-impermeable enteric coating, each push-pull unit is prevented from absorbing water in the acidic aqueous environment of the stomach. Thus, no drug is delivered. As the units enter the small intestine, the coating dissolves in this higher pH environment (pH >7). Water enters the unit, causing the osmotic push compartment to swell, and concom-

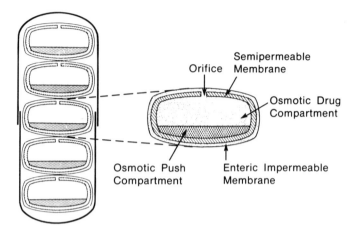

Figure 8   Cross-section of the OROS-CT colon-targeted drug delivery system.

itantly creates a flowable gel in the drug compartment. Swelling of the osmotic push compartment forces drug gel out of the orifice at a rate precisely controlled by the rate of water transport through the semipermeable membrane.

One way in which a delay period of different duration may be programmed is by incorporating a drug-free layer next to the delivery orifice, such that the nondrug formulation is delivered over the desired delay period. For treating ulcerative colitis, each push-pull unit is designed with a 3–4 hr postgastric delay to prevent drug delivery in the small intestine. Drug release begins when the unit reaches the colon. OROS-CT units can maintain a constant release rate for up to 24 hr in the colon or can deliver drug over an interval as short as 4 hr. While one single OROS-CT unit can deliver drug at any point in the colon or throughout the colon, an OROS-CT with multiple units (OROS-CTmu) provides added dispersion in time and location for drug delivery. When an OROS-CTmu releases its units in the stomach, each single unit exits as dictated by the gastric emptying physiology under conditions of a light or heavy meal (Fig. 4), arrival of the units in the colon can be expected up to 10 hr from ingestion.

Figure 9 shows the in vitro release profile of an OROS-CTmu designed to deliver 200 μg of beclomethasone dipropionate as topical treatment of the colon in ulcerative colitis. The systems did not

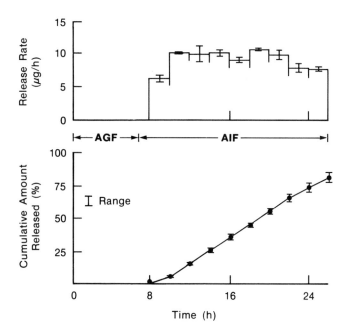

Figure 9 In vitro release rate profile of a multiunit OROS-CT delivering 200 μg of beclomethasone dipropionate for the treatment of ulcerative colitis. AGF, artificial gastric fluid; AIF, artificial intestinal fluid.

release drug in artificial gastric fluid during a period of 6 hr and exhibited a delay in release onset of about 3 hr after being transferred to artificial intestinal fluid. These systems, as programmed, continued to release for about 20 hr after that.

B. Colon-targeted (Osmet) Osmotic Pump for Clinical Research

For use in clinical research, a miniature osmotic pump (Osmet delivery system) manufactured by Alza can be changed to provide convenient controlled release of therapeutic agents into the colon. This cylindrical device (Fig. 10) has dimensions of 25 × 7 mm and a reservoir capacity of 200 μl. It is supplied empty to researchers, who fill it with a solution or suspension of the agent being studied. In the alimentary tract it imbibes water osmotically at a rate pre-

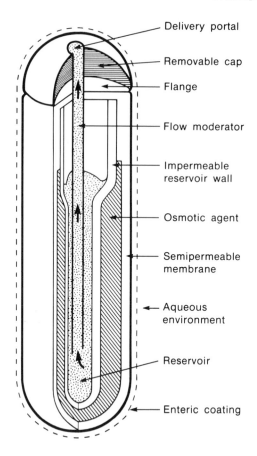

Figure 10　Cross-section of an enteric-coated, colon-targeted osmotic pump: Osmet-CT.

cisely controlled by a semipermeable membrane surrounding the drug reservoir. The imbibed water generates hydrostatic pressure inside the system, compressing the flexible, chemically inert reservoir to achieve a constant delivery of drug formulation through an orifice. The system can be designed to deliver its contents over 4–24 hr. For colonic delivery, a coating is applied to the outside of the pump that is insoluble in gastric fluid but soluble at a pH of 7 or above. Such a modified Osmet pump is designed to delay initiation of drug delivery for 2–4 hr after passing from the stomach.

The performance of an 8 hr Osmet system was proven in vitro and clinically with gamma scintigraphy (35). In vitro, the system did not release in simulated gastric fluid nor during the first 3–4 hr in simulated colonic fluid. After that, the unit delivered drug in solution at a constant rate of 20.4 µl/hr for a period of 8–10 hr (Fig. 11). Studies using gamma scintigraphy in seven healthy subjects showed that the average residence time in the empty stomach was

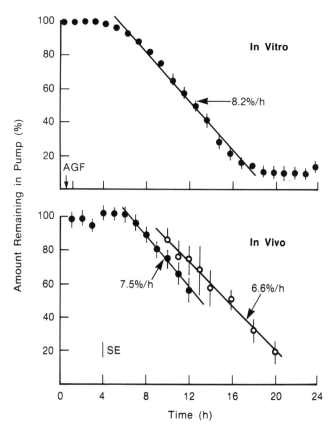

Figure 11   Delivery of drug in vitro from enteric-coated, colon-targeted Osmet-CT osmotic pumps designed to deliver 9.1% of their drug load per hour for 8 hr. Pumps were tested in simulated colonic fluid (mean ± SE; from ref. 34).

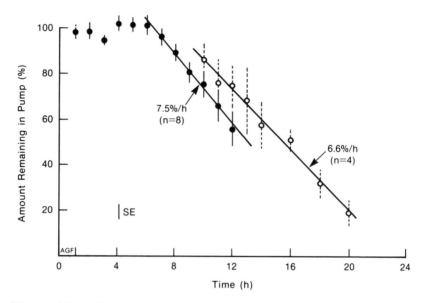

Figure 12   Delivery of drug in vivo from enteric-coated, colon-targeted osmotic pumps designed to deliver either 7.5% (n = 8) or 6.6% (n = 4) of their drug contents per hour (mean ± SE).

1.2 hr (range, 0–3 hr). Mouth-to-cecum transit time was 6.4 hr (range, 5–9 hr). The mean start-up time for the system was 5.3 hr and the rate of discharge was between 16 and 17 μl/hr. These values are based on analysis of the residue from recovered, spent pumps (Fig. 12).

The Osmet system therefore offers a safe and effective way to deliver constantly 200 μl of a medication from the cecum throughout the colon for 4–24 hr. This thoroughly studied system can serve well as a research tool to deliver drugs of different properties to the colon for local action or systemic absorption.

V. DRUG CANDIDATES FOR COLONIC TARGETING

Therapeutic agents that would benefit from colon targeting include drugs for the treatment of inflammatory bowel disease and irritable bowel syndrome. Also drugs metabolized in the upper gastrointestinal tract, hindering their systemic absorption, would be candidates for colon targeting.

A. Treatment of Inflammatory Bowel Disease

OROS-CT technology can enhance the benefits of drugs for the local treatment of ulcerative colitis and Crohn's disease that are currently administered by enema or by colon-targeted, immediate-release dosage forms (36). These drugs include salicylates such as 5-ASA and 4-ASA; corticosteroids such as beclomethasone dipropionate, tixocortol pivalate, budesonide, and prednisolone; the mast cell inhibitor cromoglycate; and the cytoprotective agent sucralfate. Their administration by OROS-CT would confer the therapeutic advantage of ensuring their controlled delivery to all colonic regions.

B. Treatment of Irritable Bowel Syndrome

Irritable bowel syndrome (IBS) has an obscure cause. It is characterized by abdominal pain and changes in bowel movements, including constipation and diarrhea, without any observed pathologic abnormalities (37). These symptoms recur at irregular intervals and stress may exacerbate them. Several studies have suggested that the syndrome is associated with an abnormality of colonic motility, exaggerated by the consumption of food (38–40). Motility-reducing agents such as calcium channel blockers, anticholinergics, and spasmolytic agents have exhibited some effectiveness in IBS, acting to minimize abnormal postprandial colonic contractions and providing symptomatic relief. Targeting motility-reducing agents to the colon may provide adequate treatment for IBS and allow for a reduction in dosage and systemic side effects.

Although the definition of IBS, the identification of drugs for its treatment, and treatment modalities are largely unspecified, the opportunity to conduct studies is great. Given the status of colon-targeted OROS-CT technology, the availability of agents, and the research tool, this area appears ripe for pharmacodynamic studies.

C. Systemic Delivery

Thus far, no delivery system has been available to provide a convenient, versatile, and effective delivery of drugs throughout the colon. Moreover, little research has been done to characterize the permeability of the colon and its potential for the absorption of proteins, peptides, and other agents. However, OSMET-CT can serve as a convenient research tool to study colonic absorption in humans before the development of dedicated solid oral dosage forms. Alza and other research teams are actively engaged in such studies, and the new information generated is expected to accelerate the development of dedicated delivery systems.

VI. CONCLUSION

Targeting of drug delivery to the colon addresses important unmet therapeutic needs for local treatment including oral drug delivery and absorption of peptides. Many questions remain regarding the causes of colonic disease and colonic permeability. However, the state of the art has progressed to the point where research tools exist to aid in the design of prototypes for the commercial development of colon-targeted products.

The OROS-CT colon-targeted systems are an extension of osmotic system technology. This basic osmotic technology has spawned highly successful commercial products such as Procardia XL. As the development and manufacturing capabilities for osmotic systems have been established, these capabilities should contribute importantly to establishing this new line of colon-targeted osmotic system products.

REFERENCES

1. C. F. Garland, A. M. Lilienfeld, A. I. Mendeloff, J. A. Markowitz, K. B. Terrell, F. C. Garland, *Gastroenterology* 81: 1115 (1981).
2. H. J. Hodgson, *Br. J. Clin. Pharmacol.* 14: 159 (1982).
3. H. D. Janowitz, *Am. J. Gastroenterol.* 82: 498 (1987).
4. J. L. A. Roth, in *Gastroenterology*, (H. L. Bockus, ed.) W. B. Saunders, Philadelphia (1964).
5. M. Friedman, L. B. Strang, *Lancet* 2: 569 (1966).
6. K. B. Klein, *Drug Ther.* 17(8): 17d (1987).
7. S. A. W. Gibson, C. McFarlan, S. Hay, G. T. MacFarlane, *Appl. Environ. Microbiol.* 55: 679 (1989).

8. G. T. MacFarlane, J. H. Cummings, S. MacFarlane, G. R. Gibson, *J. Appl. Bacteriol.* 67: 521 (1989).
9. M. Saffran, G. S. Kumar, C. Savariar, J. C. Burnham, F. Williams, and D. C. Neckers, *Science* 233: 1081 (1986).
10. M. Saffran, G. S. Kumar, D. C. Neckers, J. Pena, R. H. Jones, B. Field, *Biochem. Soc. Trans.* 18: 752 (1990).
11. P. Gruber, M. A. Longer, J. R. Robinson, *Adv. Drug Del. Rev.* 1: 1 (1987).
12. T. Nishihata, Y. Okamura, H. Inagaki, M. Sudho, A. Kamada, T. Yagi, R. Kawamori, M. Shichiri, *Int. J. Pharm.* 34: 157 (1987).
13. M. Jay, G. A. Digenis, T. S. Foster, D. R. Antonow, *Dig. Dis. Sci.* 31: 139, (1986).
14. G. M. Carlson, R. W. Rudden, C. C. Hug, P. Bass, *J. Pharmacol. Exp. Ther.* 172: 367 (1970).
15. C. F. Code, J. A. Marlett, *J. Physiol. Lond.* 246: 289 (1975).
16. R. A. Hinder, K. A. Kelly, *Am. J. Physiol.* 233: E 335 (1977).
17. J. H. Meyer, J. Dressman, A. Fink, G. Amidon, *Gastroenterology* 89: 805 (1985).
18. H. J. Smith, M. Feldman, *Gastroenterology* 91: 1452 (1986).
19. R. Khosla, L. C. Feely, S. S. Davis, *Int. J. Pharm.* 53: 107 (1989).
20. R. Khosla, S. S. Davis, *Int. J. Pharm.* 62: R9 (1990).
21. S. S. Davis, in *Topics in Pharmaceutical Sciences 1983*, (D. D. Breimer, P. Speiser, eds.) Elsevier Science P, Amsterdam (1983).
22. S. S. Davis, J. G. Hardy, M. J. Taylor, D. R. Whalley, C. G. Wilson, *Int. J. Pharm.* 21: 167 (1984).
23. J. G. Hardy, S. S. Davis, R. Khosla, C. S. Robertson, *Int. J. Pharm.* 48: 79 (1988).
24. J. Sjogren, C. Bogentoft, in *Optimization of Drug Delivery* (H. Bungaard, A. Bagger-Hansen, H. Kofod, eds.) Munksgaard, Copenhagen (1982).
25. G. Jarnerot, *Drugs* 37: 73 (1989).
26. A. Rubenstein, *Biopharmaceut. Drug Dispos.* 11: 465 (1990).
27. M. C. Rijk, R. A. van Hogezand, A. van Schraik, J. H. van Tongeren, *Scand. J. Gastroenterol.* 24: 1179 (1989).
28. C. Larsen, E. Harboe, M. Johansen, H. P. Olesen, *Pharm. Res.* 6: 995 (1989).
29. D. R. Friend, G. W. Chang, *J. Med. Chem.* 28: 51 (1985).
30. *Scrip Rep* 1504: 26 (1990).
31. P. Zeitoun, P. Brisard, U.S. Patent 4,432,966 (1984).
32. F. Martin, *Dig. Dis. Sci.* 32: 57 S (1987).
33. F. Theeuwes, G. Guittard, P. Wong, U.S. Patent 4,904,474 (1990).
34. D. Swanson, B. Barclay, P. Wong, F. Theeuwes, *Am. J. Med.* 83 (suppl. 6B): 3 (1987).
35. A. Chacko, K. F. Szaz, J. Howard, J. H. Cummings, *Gut* 31: 160 (1990).

36. L. Sutherland, *Med. Clin. North Am.* 74: 119 (1990).

37. T. P. Almy, R. I. Rothstein, *Annu. Rev. Med.* 38: 257 (1987).

38. F. Narducci, G. Bassotti, M. T. Granata, M. A. Pelli, M. Gaburri, R. Palumbo, A. Morelli, *Dig. Dis. Sci.* 31:241 (1986).

39. F. Narducci, G. Bassotti, M. Gaburri, F. Farroni, A. Morelli, *Am. J. Gastroenterol.* 80: 317 (1985).

40. A. Prior, S. R. Harris, P. J. Whorwell, *Gut* 28: 1609 (1987).

8

Colonic Delivery of Therapeutic Peptides and Proteins

Martin Mackay
Ciba-Geigy Pharmaceuticals, Horsham, West Sussex, England

Eric Tomlinson*
Somatix Therapy Corporation, Alameda, California

I. INTRODUCTION

To date, a limited number of peptides and proteins have been used therapeutically. For example, insulin, oxytocin, bradykinin, factor VIII, and calcitonin have been synthesized chemically or extracted from tissue and have been used to great effect. However, advances in cell and molecular biology have led to a much greater perception of the therapeutic value of many other peptides and proteins. Recombinant DNA technology has enabled high-level expression of this new class of therapeutic agent (1, 2) and biotechnology has permitted their large-scale production for clinical use (3). Examples of recombinant polypeptide and protein drugs already commercially available include human insulin (approved almost a decade ago), tissue plasminogen activator (TPA) (4), human growth hormone (hGH), interleukin-2 (IL-2) (5–7), and human erythropoietin (8). Others, such as hirudin (9), human interferon (10), human tumor necrosis factor (TNF) (11), interleukin-1 (IL-1) (12), and he-

---

* *Present affiliation*: GeneMedicine, Inc., Houston, Texas

mopoietic growth factors (13), are undergoing clinical trials. In their 1990 annual survey, the Pharmaceutical Manufacturers Association (PMA) stated that more than 100 biotechnologically produced therapeutic agents are undergoing clinical evaluation and 11 have already gained approval from the United States Food and Drug Administration (FDA). The peptides and proteins in trial include a vast range of molecules with markedly different physical and chemical properties and physiological actions (see Table 1). Molecular weight and conformation vary enormously: from somatostatin, a cyclic 14 amino acid regulatory hormone with a molecular weight of 1638, to cardiovascular-active peptides such as TPA, which are 567 amino acid agents with a molecular weight of 59,050. Posttranslational modifications also differ: proteins such as interferon-β1, interferon-β2, and erythropoietin are naturally glycosylated while many (for example, insulin, albumin, and interferon-α) are not. In addition, the ranges of biological effects and modes of action are considerable.

It is a major challenge to scientists within academia and the pharmaceutical industry to ensure that convenient systems are devised for the delivery of peptide and protein drugs. These systems will have to deliver peptide and protein drugs in a safe and pharmacologically relevant manner, and this subject has gained attention recently (14–16). In addition, the route of administration will have to be acceptable to the patient and physician given any particular indication. To date, almost all therapeutic peptides and proteins are administered by injection and most will have to be delivered in this manner, at least for the foreseeable future. However, it may be expected that for the treatment of chronic disease in non-life-threatening situations such parenteral administration will lead to poor patient compliance and thus to restricted use. Great efforts are being directed towards the effective use of alternative routes of administration. "Opportunity windows" have been demonstrated for peptide and protein delivery (17). For example, there is an extensive literature on the potential of using non-injectable routes such as nasal, buccal, rectal, and vaginal (18, 19) administration, and success has been achieved with an intranasal formulation of salmon calcitonin now on the market.

It has recently been recognized that specific regions of the gastrointestinal (GI) tract may offer potential in terms of macromolec-

Table 1 Examples and Application of Peptides and Proteins in Clinical Use or Undergoing Clinical Trial

Peptide or protein drug	Application
Interleukin-2	Renal carcinoma
Interleukin-1	Carcinoma
Tissue necrosis factor	Carcinoma
Colony-stimulating factor	Carcinoma
Epidermal growth factor	Wound healing
Transforming growth factors	Wound healing
Fibroblast growth factor	Wound healing
Insulin-like growth factors	Wound healing
Hirudin	Fibrinolytic
Tissue plasminogen activator	Fibrinolytic
Urokinase	Fibrinolytic
Streptokinase	Fibrinolytic
Erythropoietin	Erythropoiesis stimulation
Factor VIII	Hemophilia
Factor IX	Christmas disease
Insulin	Glucose regulation
Somatostatin	Glucose regulation
Proinsulin	Glucose regulation
Macrophage-inhibiting factor	Immunoregulation
Macrophage-activating factor	Immunoregulation
Muramyl dipeptide	Immunoregulation
Colony-stimulating factor	Immunoregulation
Interferons	Immunoregulation
Glucocerebrosidase	Gaucher's disease
Calcitonin	Bone disease
Oxytocin	Labor induction
Growth hormone	Dwarfism
α-1 Antitrypsin (AAT)	AAT deficiency
Superoxide dismutase	Respiratory disorders

ular absorption. The protease profile of the GI tract would suggest that some regions have reduced levels of degradative enzymes and would be attractive for peptide and protein delivery. This chapter will address the progress made in terms of peptide and protein delivery to, and absorption from, one such region: the colon.

II. THE COLONIC ENVIRONMENT: ISSUES FOR PEPTIDE AND PROTEIN DRUGS

The human colon is a versatile organ. It serves many functions, the most important being the maintenance of fluid and electrolyte balance, and facilitation of the processing of waste material for the body. In humans, between 500 and 1500 g of semiliquid material normally enters the colon per day. Most of the fluid is resorbed and an output of 100 g of feces per day results. Both the structure and function of the mammalian colon have been extensively reviewed (20–23) although, to date, there is a paucity of literature on correlations between structure/function and drug transport across the membrane.

The current dogma suggests that most types of drugs are poorly absorbed from the colon. However, reports have shown that the absorption of some drugs across different regions of the GI tract is equivalent. Brockmeier *et al.* (24) showed that the absorption of glibenclamide was similar whether it was administered to the stomach, duodenum, or colon. Given the more favorable environment of the colon compared to the stomach and small intestine in terms of proteolytic degradation, it is no wonder that the role of the colon in terms of peptide and protein drug delivery has lately been revisited. Cummings and co-workers (25) have recently shown that the human colon contains fewer contents than was once believed. The total amount of content in 46 adults was 222 ± 21 g (wet) with only 93 g in the cecum and ascending regions. In addition, other studies have increased our knowledge of the barrier function of the colon. The development of *in vitro* methods (see Chapter 3), for example primary colonic epithelial cultures (26, 27), cell lines derived from colonic adenocarcinomas (28–30), and cell lines derived from rectal adenocarcinomas (31–34), has given greater insight into the function of colonic epithelial cells. Cell cultures are also being used to study the properties of the colon, such as transport ability.

The colonic environment presents a complex mixture of microorganisms and degradative enzymes to any drug. More than 400 different species of bacteria reside in the colon and they are represented by both aerobic and anaerobic species, although the anaerobes predominate (35). They number approximately 10^{11}/ml and this results in almost one-third of the dry weight of feces consisting of bacteria (36). Colonic bacteria secrete a plethora of enzymes, mainly involved in the fermentation of a variety of substrates for energy purposes. Some of these enzymes have been used as "triggers" for colon-specific drug delivery. In addition to bacterially produced enzymes, it is believed that a small amount of residual pancreatic proteases are present. Although there are few data on the relative amounts of pancreatic enzymes in the small intestine and colon, what evidence is available suggests that the concentration is much reduced in the colon. Gibson and co-workers (37) reported on the significance of microflora in proteolysis in the human colon. They showed that proteolytic activity was significantly greater, by an order of magnitude, in small intestinal effluent than in feces. Moreover, fecal proteolysis was qualitatively different from ileal proteolysis, as evidenced by the range of proteins hydrolyzed and the susceptibility of the enzymes to protease inhibitors such as chymostatin.

The active transport properties of the colon have been reviewed recently (23). Transport pathways in the colon provide for rapid and specific active bidirectional transport of ions across the epithelium. Unlike in the small intestine, there are no documented active transporters for organic nutrients in the mature organ. The lack of such transporters may limit the scope for drug design with respect to mediated transport across the epithelial barrier and, therefore, drug absorption at this site is a consequence of the general properties and features of the colon. However, receptors for peptides and proteins such as insulin (38) and insulin-like growth factor I (IGF-I) (39) have been found on the apical membrane of rabbit proximal colon epithelial cells. Whether this phenomenon can be exploited to deliver peptide and protein drugs specifically remains to be seen, and a detailed study is needed to investigate the physiological role for such receptors and the capacity for transepithelial transport of drugs. The general properties of the colon that may be of significance in terms of peptide and protein absorption include the following:

Lack of proteolytic enzymes,
The luminal milieu has been shown to be suitable for prodrug metabolism, as with salicylazosulfapyridine, or offer an environment suitable for drug delivery or site-specific drug release,

Studies with absorption enhancers have shown that the colon is a
   suitable site for such approaches,
The transmucosal and membrane potential differences may be of
   significance in the absorption of ionized or ionizable drugs,
The bulk water absorption in this region of the intestine may provide scope for solvent drag.

These barrier function properties will influence the choice of the
colon as a region for peptide and protein absorption after oral administration; examples are cited below.

III. PEPTIDE ABSORPTION

It is unfortunate that the physical and chemical properties of proteins do not make them ideal candidates for oral administration.
Their size, often complex secondary, tertiary, and even quaternary
structure, and their surface characteristics give rise to this apparent poor candidacy. For example, for a drug to be administered
orally and have a therapeutic effect, it must withstand the following hostilities: chemical and enzymic degradation during transit,
the mucosal barrier, metabolism during transepithelial transport,
and "first-pass" hepatic metabolism.

The GI absorption of drugs has often been reviewed (40–42). It
has been known for several decades that peptides and proteins do
cross the GI tract, albeit in low amounts (43–53). These peptides
and proteins include horseradish peroxidase (HRP), chymotrypsin,
elastase, *Clostridium botulinum* type A toxin, a variety of food anti-
gens, and insulin. It is clear that more than one mechanism exists
by which intact peptides enter the blood compartment *via* the GI
tract and these include both passive and specific pathways.
Whether these normal physiological processes can be exploited for
the delivery of therapeutic peptides and proteins across the colon
remains unanswered. It is generally accepted that these processes
only enable the transport of minute quantities of macromolecules
and therefore would be insufficient to achieve therapeutic levels of
most peptides or proteins.

To date, few studies have been reported in which a peptide or
protein has been administered intracolonically in the absence of
protease inhibitors or enhancers. Atchison *et al.* (54) investigated
the colonic absorption of radiolabeled insulin using noneverted

sacs of rat colon. They measured the percentage of intraluminal insulin degradation and the transport of insulin into the surrounding media. They showed that transepithelial flux of insulin was consistently less than 0.3% of the administered dose. In addition, significant degradation of insulin (64%) occurred on the luminal side within 15 min of exposure.

The effects of intracolonically administered human insulin, and human insulin–diethylaminoethyl (DEAE)–dextran complex entrapped in liposomes, on blood glucose in rats has been reported (55). In both cases, a hypoglycemic response was noted that lasted for several hours. No indication of relative bioavailability was given and it can only be presumed that insulin was transported across the colon. Lundin and Vilhardt (56) measured the absorption of a vasopressin analog, 1-deamino-8-D-arginine-vasopressin (dDAVP) from different regions of the GI tract of rats: stomach, duodenum, midpart of the ileum, the ileocecal junction, and the midpart of the colon. They showed that absorption of intact dDAVP occurred from each site and was rapid, with peak plasma levels after 10–20 min. Less absorption was observed across the stomach and colon. The authors speculated that low surface area associated with these regions compared to the small intestine was the reason.

The few *in vivo* studies that have been performed have neither measured bioavailability nor determined the mechanism of absorption. Moreover, there is a need to study the absorption of therapeutic peptides and proteins in humans. Recently, human cell lines have been used as intestinal models to investigate the transport of drugs. These cells have been grown as monolayers on filters and been shown to display many characteristics of differentiated intestinal epithelial cells (29, 57). For example, an adenocarcinoma cell line, Caco-2, derived from a human colon (58), was used to study the transport of HRP (29). Low concentrations of apically applied HRP were seen in the basolateral chamber (0.0001%). Using the same colonic cell line, the absorption profiles of arginine-vasopressin (AVP) and dDAVP (59) were determined. Both these hydrophilic peptides displayed linear transport kinetics and were absorbed at very slow rates by a nonsaturable, nonpolar process across the Caco-2 monolayers. Neither peptide was degraded after a 4 hr incubation in the apical medium of the cells.

The use of such cell models to investigate peptide transport is clearly at an initial stage, but early results indicate their potential usefulness.

It is clear that the few experimental data that exist show that therapeutic peptides and proteins are poorly absorbed from the GI tract. Although the colon may be a preferable site for peptide delivery because of the relative lack of proteases, this fact alone does not lead to effective bioavailability. As a result, various strategies have been applied to elevate absorption levels and include protection against presystemic degradation, chemical derivatization of the drug (prodrugs), and coadministration of so-called absorption enhancers. In a few instances these have been used successfully to promote the absorption of peptides and proteins.

IV. ENHANCEMENT

Attempts to increase drug absorption from the GI tract by the coadministration of enhancers started in earnest in the early 1960s. Surfactants and chelating agents were used but were shown to be impractical due to the extensive epithelial cell damage that resulted from their use. Recently, much work has focused on increasing the low amounts normally associated with peptide and protein availability. Attempts have been made to circumvent the difficulties associated with the low bioavailability of the oral route, with varying degrees of success. One reason for slower than expected progress has been the lack of understanding of the nature and complexity of the reasons for poor bioavailability. The belief that by simply protecting peptides from proteolytic degradation one could solve low availability after oral administration was naive and did not, for example, take into account the barrier function of the epithelial membrane and associated layers (60–62). The coadministration of protease inhibitors with insulin demonstrated little increase in absorption (45, 63). However, more recent reports have shown that by protecting insulin against luminal degradation, absorption is increased (64) and probably reflects the improvement in the inhibitors used.

Attempts to bridge the epithelial membrane using enhancers, accelerators, or promoters (thereby allowing increased peptide absorption) has often been rather primitive, resulting in pathologic damage to the epithelium. However, progress has been made and should provide a platform for more rational approaches for at least some therapies. Comprehensive reviews are presently available with respect to this field (65–67). Various adjuvants have been used over several decades to enhance the absorption of macromolecules across different regions of the GI tract, including nonsteroidal anti-inflammatory drugs (NSAIDs), for example, salicylates (68) and diclofenac (69), water-oil-water emulsions (70), surfactants (71), bile salts (72), surfactant–lipid mixed micelles (73–76), and additive promotion effects of azone and fusogenic fatty acids (77).

Nishihata and co-workers (78) compared the enhancing action of sodium salicylate, disodium ethylenediaminetetraaecetic acid (EDTA), diethyl ethoxymethylenemalonate (DEEMM), diethyl maleate (DEM), and trifluoperazine (TFP) on colonic absorption of [Asu^{1,7}]-eel calcitonin in rats. The absorption of [Asu^{1,7}]-eel calcitonin in the absence of enhancers was thought to be predominantly *via* the paracellular route but at "a very low rate." The authors suggested that EDTA and TFP enhanced paracellular absorption. An interesting finding was that the enhancing action of salicylate and DEEMM at low concentrations was *via* the transcellular route whereas at higher concentrations it occurred predominantly *via* a paracellular route. DEM did not enhance colonic absorption of [Asu^{1,7}]-eel calcitonin.

Ziv *et al.* (79, 80) demonstrated that insulin was absorbed from the colon of rats in the presence of sodium cholate, extracts of ox bile, or sodium deoxycholate. When similar experiments were performed in the small intestine, absorption was only seen when a protease inhibitor was coadministered. No protease inhibitor was necessary for colonic absorption, suggesting the scarcity of proteases in the colon. Once again bioavailability was not determined and the biological activity results suggested that only small amounts of insulin were transported. A rat colonic sac model has been used to show the enhancing properties of phenothiazines, DEM, and EDTA (81). Although no peptide was tested, enhanced absorption of the macromolecule inulin (molecular weight 5000) was demonstrated. Absorption of other macromolecules from the colon of rats has been shown using the nonionic surfactant Cetomacrogol 1000 (82). Heparin was administered by microenema and thrombin time was measured. Prolongation of thrombin time was demonstrated 1 hr after administration and, in most cases cited, rendered the plasma uncoagulable. No indication of bioavailability was given. Although the mechanism of enhancement was not elucidated, preliminary toxicity tests showed no adverse histologic effects.

Muranishi has shown that enhancement of macromolecule absorption, for example, heparin, by mixed micelles is more effective in the colon than in the small intestine, resulting in higher and faster peak plasma levels. Moreover, the minimum concentration of mixed micelles needed to potentiate enhancement was significantly lower (25%) (67). The increased enhancement of macromolecule absorption afforded by the so called "ufasomes," which are closed vesicles with bilayer-like structures, was also greater in the colon than in the small intestine (83).

A plethora of explanations are given for increased intestinal permeability due to absorption enhancers. They include extracellular actions, such as binding of mucosal Ca^{2+}; membrane effects, for example, increasing membrane fluidity and surfactant effects; and intracellular events, such as reduction in glutathione levels (84). In addition, different enhancer "classes" may have specific effects. A number of mechanisms have been postulated for the enhancing effect of salicylates, including reduction of the mucous layer (85) and increase in lipid bilayer fluidity (86). Bile salts have been shown to affect the mucosal barrier in different ways: alteration of the physicochemical properties of the mucus, solubilization of mucous components; and even stimulation of mucous secretion (87, 88). Many explanations have been offered for the enhancing mechanism of mixed micelles. They include mucous effects and transcellular mechanisms (67). Awazu and co-workers examined the effects of sodium caprate and sodium caprylate on brush border membrane vesicles (BBMVs) from rat colon (89, 90). They showed that increased permeation was due to permeability *via* the transcellular route caused by membrane perturbation. It is unfortunate that these enhancers were not tested to determine whether they increased absorption of peptides *in vivo*.

Although the enhancing agents described undoubtedly increase absorption, there have been few attempts to determine the predominant mechanism of action or, for that matter, the toxicologic effects resulting from chronic application. It is now clear that many enhancers merely perturb the epithelial membrane, albeit transiently, such that endogenous or food-derived molecules in close proximity to the drug and enhancer may also cross into the vascular compartment. The extent of this "leakiness" is dependent on the enhancer and the concentration. The lack of "specificity" or selectivity of many enhancers and the inevitable uptake of bystander molecules in the vicinity of the drug/enhancer dosage form is a major concern. The consequence of this in terms of toxicity can only be assessed in chronic studies, of which there are few. However, one could predict that the absorption of endogenous intestinal enzymes would lead to clinical pathologic findings. A greater understanding of cell physiology and biophysical forces should lead to the development of "safe" enhancement systems that do not cause pathologic damage to the epithelia and are specific for the drug to be transported.

V. COLONIC DELIVERY

A number of ways to deliver drugs specifically to the colon have been cited (35). Colonic delivery can be achieved with low "molecular" weight drugs by making prodrugs that survive passage through the small intestine where they are not absorbed and whose active moiety is released by enzymes specifically produced in the colon. For example, sulfasalazine (salicylazosulphapyridine) is poorly absorbed across the small intestine and is reduced to 5-amino salicylic acid and sulfapyridine by azoreductase produced by anaerobic bacteria in the colon (91, 92). The cleavage of sulfasalazine is markedly reduced in patients taking antibiotics or who have had their colon removed. A colon-specific drug delivery system has been described based on steroid glycosides and the glycosidases of colonic bacteria (93–95). Steroid glycosides are poorly absorbed from the small intestine. However, on reaching the colon, bacterial glycosidases release free steroid, for example, dexamethasone, which is absorbed from the colon. Whether the prodrug approach can be used for the delivery of therapeutic peptides and proteins remains to be seen. However, enzymes produced by colonic bacteria have been used to degrade delivery devices.

Exploitation of enzymes naturally produced by colonic bacteria is common to many colon-specific drug release concepts. In terms of peptides and proteins, devices have been invented that specifically release the drug in the colon rather than being based on a prodrug principle. Saffran and co-workers have developed a system in which insulin or vasopressin is encapsulated in a gelatin capsule coated with an impermeable polymer (96, 97). The coat is resistant to the degradative processes of the stomach and small intestine and is prepared using azo functional cross-linking agents based on divinylazobenzene. In a similar reaction to sulfasalazine, the azoaromatic bonds ($R-C_6H_4-N=N-C_6H_4-R$) are cleaved by reduction to form a pair of aromatic amines ($R-C_6H_4-NH_2$ + $H_2N-C_6H_4-R$) by bacterial azoreductases, thus breaching the capsule and releasing the drug specifically in the colon. In the case of insulin, azopolymer-coated capsules were given orally to diabetic dogs and a hypoglycemic response resulted (blood glucose was reduced from 4 mg/ml to 2.9 mg/ml) 3 hr after oral administration. With vasopressin, azopolymer-coated capsules were placed in the stomach of an anesthetized rat and antidiuresis measured. Maximum antidiuresis resulted 3 hr following administration of the capsule. Both the insulin and vasopressin data were subject to a high degree of variability, probably due to the irreproducibility of coating the capsules. Other workers have attempted to devise more reliable systems.

Kopecek and co-workers (98–100) have developed novel types of hydrogel capsules, based on acrylic acid, N,N-dimethylacrylamide, and N-tert-butylacrylamide cross-linked with 4,4'-di(methacryloylamino)azobenzene. These hydrogels do not swell significantly in the stomach. However, in transit through the small intestine, swelling increases due to increased pH. In the colon, the degree of swelling renders the cross-links accessible to bacterial azoreductases and therefore mediates breakdown of the hydrogel and release of the drug. The capsules have been tested *in vitro* and *in vivo* in degradation experiments and display a high degree of reproducibility. However, few data have been published demonstrating the effectiveness of such a system for colonic release of peptide or protein drugs, and one can only speculate on their utility.

Davies *et al.* (101) have patented a capsule for targeted delivery of peptides to the colon. The capsule is coated with a 60–150 μm thick layer of the polyacrylic polymer, Eudragit S, a commercially available anionic polymer. The patent claims that the coating is insoluble in the gastric and small intestinal milieu below a pH of 7.0 but soluble in the colonic environment. Insulin, porcine calcitonin, and hGH have been encapsulated as formulations also containing absorption enhancers such as sodium 5-methoxy salicylate, wetting agents, and surfactants. Apart from concerns relating to the enhancers used, the reproducibility of such as system for colonic delivery in humans remains to be demonstrated. In another study, Touitou and Rubinstein (102) used a similar type of capsule, with porcine insulin. The formulation also contained sodium lauratecetyl alcohol (2:8) and a hypoglycemic response was demonstrated in rats, although relative bioavailability was stated to be "low." pH-Sensitive polymers have also been reported and rely on the assumption that the proximal colon has a pH of 8.0–8.4; the polymer only dissolves in this pH range. However, the pH in other regions of the GI tract can rise to this level, rendering these polymers unreliable for the release of drug only in the colon (103).

It is apparent that specific release of peptide and protein drugs in the colon is possible. The possibility of polymeric delivery systems that degrade when exposed to enzymes produced by the colonic flora has been shown and awaits appropriate peptide formulation for clinical trials.

VI. CONCLUSIONS

A great deal of research is being aimed at achieving effective administration of therapeutic peptides and protein by nonparenteral routes. Opportunities have been identified that could lead to more convenient delivery systems for this class of drug and the prospect of marketed products is appealing. The colon is clearly a region of the GI tract worthy of attention, given the apparent paucity of enzymes associated with proteolytic degradation. Attempts to achieve therapeutic levels of peptide and protein drugs in the vascular compartment following intracolonic administration have mostly relied on absorption enhancers. The utility of such enhancers for chronic administration of drug is doubtful, given the epithelial damage that many have been shown to cause. However, the need for "safe" methods to achieve effective peptide absorption is unquestionable and would be of great benefit to patients and physicians. One can envisage that judicious efforts by pharmaceutical technologists and molecular cell biologists will lead to the targeting of peptide and protein drugs to specific regions of the GI tract, such as the colon, and lead to improved therapies.

ACKNOWLEDGMENTS

Many thanks to Peter Goddard, John Hastewell, and Alan Steward for criticism of the original manuscript, and to Christine Malin for its preparation.

REFERENCES

1. D. T. Liu, N. Goldman, F. Gates III, *Lymphokine Res.* 5: S189–S192 (1986).
2. S. K. Sharma, *Ad Drug Del. Rev.* 4: 87–112 (1989).
3. R. M. Baum, *Chem. Eng. News* 65: 11–32 (1987).
4. D. Collen, *Circulation* 72: 18–23 (1985).
5. S. A. Rosenberg, M. T. Lotze, L. M. Muul, A. E. Chang, F. P. Avis, S. Leitman, W. M. Linehan, C. N. Robertson, R. E. Lee, J. T. Rubin, C. A. Seipp, C. G. Simpson, D. E. White. *N. Engl. J. Med.* 316: 891–897 (1987).
6. G. Stoter, E. Shiloni, S. Aamdal, F. J. Cleton, S. Iacobelli, J. Th. Bijman, P. Palmer, C. R. Franks, S. Rodenhuis *Eur. J. Cancer Clin. Oncol.* 25: S41–S43 (1989).
7. S. Negrier, T. Philip, G. Stoter, S. D. Fossa, S. Janssen, A. Iacone, F. S. Cleton, O. Eremin, L. Israel, C. Jasmin, C. Rugarli, H. V. D. Masse, N. Thatcher, M. Symann, H. H. Bartsch, L. Bergman, J. T. Bijman, P. A. Palmer, C. R. Franks, *Eur. J. Cancer Clin. Oncol.* 25: S21–S28 (1989).
8. A. Erslev, *N. Engl. J. Med.* 316: 101–103 (1987).
9. T. Lindhout, R. Blezer, H. C. Hemker, *Thromb. Haemost.* 3:464–468 (1990).
10. T. C. Merigan, *N. Engl. J. Med.* 318: 1458–1460 (1988).
11. P. B. Chapman, T. J. Lester, E. S. Casper, J. L. Gabrilove, G. Y. Wong, S. J. Kempin, P. J. Gold, S. Welt, R. S. Warren, H. F. Starnes, S. A. Sherwin, L. J. Old, H. F. Oettgen, *J. Clin. Oncol.* 5: 1942–1951 (1987).
12. C. A. Dinarello, J. G. Cannon, J. W. Mier, H. A. Bernheim, G. A. Lopreste, D. L. Lynn, R. N. Love, A. C. Webb, P. E. Auron, R. C. Reuben, A. Rich, S. M. Wolff, S. D. Putney, *J. Clin. Invest.* 77: 1734 (1986).
13. G. Morstyn, A. W. Burgess, *Cancer Res.* 48: 5624–5637 (1988).
14. L. M. Sanders, *Eur. J. Drug Metab. Pharmacokinet.* 15: 95–102 (1990).
15. E. Tomlinson, *Adv. Drug Del. Rev.* 1: 87–198 (1987).
16. E. Tomlinson, in *Novel Drug Delivery and its Therapeutic Application* (L. F. Prescott, and W. S. Nimmo, eds.) John Wiley and Sons, Chichester (1989), p 245.

17. M. Mackay, *Biotechnol. Genet. Eng. Rev.* 8: 251–278 (1990).
18. D. A. Eppstein and J. P. Longenecker, *CRC Crit. Rev. Ther. Drug Carrier Syst.* 5: 99–139 (1988).
19. J. C. Verhoef, H. E. Boddé, A. G. de Boer, J. A. Bouwstra, H. E. Junginger, F. W. H. M. Merkus, D. D. Breimer, *Eur. J. Drug Metab. Pharmacokinet.* 15: 83–93 (1990).
20. C. M. Fenoglio-Preiser, P. E. Lantz, M. Davis, M. B. Listrom, F. O. Rike, in *Gastrointestinal Pathology, An Atlas and Text*, Raven Press, New York (1989).
21. L. Bustos-Fernandez, in *Colon Structure and Function*, Plenum Medical Co., London, (1982).
22. S. F. Phillips, *Scand. J. Gastroenterol.*, 93, 1–12 (1984).
23. J. Hastewell, I. Williamson, M. Mackay, *Adv. Drug Delivery Rev.* (in press).
24. D. Brockmeier, H-G. Grigoleit, H. Leonhardt, *Eur. J. Clin. Pharmacol.* 29: 193 (1985).
25. J. H. Cummings, J. G. Banwell, I. Segal, N. Coleman, H. N. Englyst, G. T. Macfarlane, *Gastroenterology* 98: A408 (1990).
26. A. Quaroni, *Gastroenterology* 96: 535–536 (1989).
27. P. R. Gibson, E. van de Pol, L. E. Maxwell, A. Gabriel, W. F. Doe, *Gastroenterology* 96: 283–291 (1989).
28. C. Augeron, J. J. Maoret, C. L. Laboisee, E. Grasset, in *Ion-Gradient-Coupled Transport* (F. Alvarado, C. H. van Os, eds.) Elsevier, Amsterdam (1986), p 363.
29. G. Wilson, I. F. Hassan, C. J. Dix, R. Shah, M. W. Mackay, P. Artursson, *J. Cont. Rel.* 11: 25–40 (1990).
30. K. Dharmsathaphorn, J. A. McRoberts, K. G. Mandel, L. D. Tisdale, H. Masui, *Am. J. Physiol.* 246: 204–208 (1984).
31. S. C. Kirkland, *Can. Res.* 45: 3790–3795 (1985).
32. S. C. Kirkland, *Differentiation* 33: 148–155 (1986).
33. S. C. Kirkland, *J. Cell. Sci.* 91: 615–621 (1988).
34. S. C. Kirkland, I. G. Bailey, *Br. J. Cancer* 53: 779–785 (1986).
35. A. Rubinstein, *Biopharm. Drug Dispos.* 11: 465–475 (1990).
36. G. L. Simon, S. L. Gorbach, *Bacteriology of the colon*, in *Colon Structure and Function* (L. Bustos-Fernandez, ed.) Plenum Medical Co., London, (1983), p 103.
37. S. A. W. Gibson, C. McFarlan, S. Hay, G. T. Macfarlane, *Appl. Environ. Micro.* 55: 679–683 (1989).
38. D. J. Pillion, V. Ganapathy, F. H. Leibach, *J. Biol. Chem.* 10: 5244–5247 (1985).
39. D. J. Pillion, J. F. Haskell, J. A. Atchison, V. Ganapathy, F. H. Leibach, *Am. Phys. Soc.* 257: E27–E34 (1989).
40. M. J. Jackson, in *Physiology of the Gastrointestinal Tract*, 2nd ed. (L. R. Johnson, ed.) Raven Press, New York (1987) p 1597.

41. D. T. O'Hagan, K. J. Palin, S. S. Davis, *CRC Crit. Rev. Ther. Drug Carrier Systems* 4: 197–220 (1987).
42. T. T. Kararli, *CRC Crit. Rev. Ther. Drug Carrier Systems* 6: 39–86 (1989).
43. S. J. Wilson, M. Walzer, *Am. J. Disabled Child.* 50: 49 (1935).
44. A. J. May, B. C. Whaler, *Br. J. Exp. Pathol.* 39: 307–316 (1958).
45. E. Danforth, R. D. Moore, *Endocrinology* 65: 118–126 (1959).
46. A. L. Warshaw, W. A. Walker, R. Cornell, K. J. Isselbacher, *Lab. Invest.* 25: 675 (1971).
47. A. L. Warshaw, W. A. Walker, *Surgery* 76: 495 (1974).
48. A. L. Warshaw, C. A. Bellini, W. A. Walker, *Am. J. Surg.* 133: 55 (1977).
49. A. L. Warshaw, W. A. Walker, K. J. Isselbacher, *Gastroenterology* 66: 987 (1974).
50. S. P. Galant, *Clin. Pediatr.* 115: 731 (1976).
51. J. N. Udall, W. A. Walker, *J. Pediatr. Gastroenterol. Nutr.* 1: 295–301 (1982).
52. M. L. G. Gardner, D. Wood, *Biochem. Soc. Trans.* 17: 934–937 (1989).
53. A. Pusztai, *Adv. Drug Del. Rev.* 3: 215–228 (1989).
54. J. A. Atchison, W. E. Grizzle, D. J. Pillion, *J. Pharm. Exp. Ther.* 248: 567–572 (1989).
55. A. Manosroi, K. H. Bauer, *Drug Dev. Ind. Pharm.* 16 (9): 1521–1538 (1990).
56. S. Lundin, H. Vilhardt, *Acta Endocrinol.* 112: 457–460 (1986).
57. I. J. Hidalgo, R. J. Raub, R. T. Borchardt, *Gastroenterology* 96: 736–749 (1989).
58. J. Fogh, W. C. Wright, J. D. Loveless, *J. Natl. Cancer. Inst.* 58: 209–214 (1977).
59. S. Lundin, P. Artursson, *Int. J. Pharm.* 64: 181–186 (1990).
60. G. von Heijne, *Curr. Top. Membranes Transport.* 24: 151–179 (1985).
61. V. Vetvicka, L. Fornúsek, *CRC Crit. Rev. Ther. Drug Carrier Systems* 5: 141–170 (1988).
62. R. E. Kleinman, W. A. Walker, *Acta Paediatr. Scand. Suppl.* 351: 34–37 (1989).
63. Y. W. Inouye, H. M. Vars, *Surg. Forum* 13: 316 (1962).
64. S. Fujii, T. Yokoyama, K. Ikegawa, F. Sato, N. Yokoo, *J. Pharm. Pharmacol.* 37: 545–549 (1985).
65. J. A. Fix, *J. Cont. Rel.* 6: 151–156 (1987).
66. E. J. van Hoogdalem, A. G. de Boer, D. D. Breimer, *Pharm Ther.* 44: 407–443 (1989).
67. S. Muranishi, *CRC Crit. Rev. Ther. Drug Carrier Systems* 7 (1): 1–33 (1990).
68. G. E. Peters, L. E. F. Hutchinson, R. Hyde, C. McMartin, S. B. Metcalfe, *J. Pharm. Sci.* 76: 857–861 (1987).

THERAPEUTIC PEPTIDE AND PROTEIN DELIVERY 175

69. H. Yaginuma, Y. Isoda, Y. Wada, S. Itoh, M. Yamazaki, A. Kamada, H. Shimazu, I. Makita, *Chem. Pharm. Bull.* 30: 1073–1076 (1982).
70. M. Shirchiri, R. Kawamori, Y. I. Goriya, M. Kikuchi, Y. Yamasaki, Y. Shigeta, H. Abe, *Acta Diabetol. Lett.* 15: 175–183 (1978).
71. J. A. Galloway, M. A. Root, *Diabetes* 21: 637–648 (1972).
72. M. Matsumura, S. Saito, *Endocrinol. Jpn.* 36: 15–21 (1989).
73. S. Muranishi, *Pharm. Res.* 2: 108–118 (1985).
74. H. Yoshikawa, K. Takada, S. Muranishi, Y. Satoh, N. Naruse, *J. Pharmacobiodyn.* 7: 59–62 (1984).
75. H. Yoshikawa, K. Takada, Y. Satoh, N. Naruse, S. Muranishi, *Pharm. Res.* 2: 249–250 (1985).
76. H. Yoshikawa, K. Takada, S. Muranishi, *Chem. Pharm. Bull.* 34: 4382–4384 (1986).
77. H. Fukui, M. Murakami, H. Yoshikawa, K. Takada, S. Muranishi, *Int. J. Pharm.* 31: 239–246 (1986).
78. T. Nishihata, M. Masatoshi, H. Takahata, A. Kamada, *Int. J. Pharmaceut.* 33: 89–97 (1986).
79. E. Ziv, M. Kidron, E. M. Berry, H. Bar-on, *Life Sci.* 29: 803–809 (1981).
80. E. Ziv Y. Kleinman, H. Bar-on, M. Kidron, in *Lessons From Animal Diabetes* (E. Shafir, ed.) John Libbey, London (1984) p 642.
81. T. Suzuka, A. Furuya, A. Kamada, T. Nishihata, *J. Pharmacobio-Dyn.* 10: 63–71 (1987).
82. M. Kidron, A. Eldor, D. Lichtenberg, E. Touitou, E. Ziv, and H. Bar-on, *Thromb. Res.* 16: 833–835 (1979).
83. M. Murakami, H. Yoshikawa, K. Takada, S. Muranishi, *Pharm. Res.* 3: 35–39 (1986).
84. A. G. de Boer, E. J. van Hoogdalem, and D. D. Breimer, *Eur. J. Drug Metab. Pharmacokinet.* 15: 155–157 (1990).
85. P. Sithigorngul, P. Burton, T. Nishihata, and L. Caldwell, *Life Sci.* 33: 1025–1032 (1983).
86. H. Kajii, T. Horie, M. Hayashi, S. Awazu, *J. Pharm. Sci.* 75: 475–478 (1986).
87. J. J. Rafter, V. W. S. Eng, R. Furrer, A. Medline, and W. R. Bruce, *Gut* 27: 1320–1329 (1986).
88. D. A. Whitmore, L. G. Brooks, K. P. Wheeler, *J. Pharm. Pharmacol.* 31: 277–283 (1979).
89. T. Sawada, M. Tomita, M. Hayashi, S. Awazu, *J. Pharmacobio. Dyn.* 12: 634–639 (1989).
90. M. Tomita, M. Hayashi, T. Horie, T. Ishizawa, S. Awazu, *Pharm. Res.* 5 (12): 786–789 (1988).
91. U. Klotz, *Clin. Pharmacokinet.* 10: 285–302 (1985).
92. J. P. Brown, G. V. McGarraugh, T. M. Parkinson, R. E. Wingard, Jr., and A. B. Onderdonk, *J. Med. Chem.* 26: 1300–1306 (1983).

93. D. R. Friend, G. W. Chang, *J. Med. Chem.* 27: 261–266 (1984).

94. D. R. Friend, G. W. Chang, *J. Med. Chem.* 28: 51–57 (1985).

95. T. N. Tozer, *Proc. Symp. Control. Rel. Bioact. Mater.* 17: 126–127 (1990).

96. M. Saffran, G. S. Kumar, C. Savariar, J. C. Burnham, F. Williams, D. C. Neckers, *Science* 233: 1081–1084 (1986).

97. M. Saffran, G. S. Kumar, D. C. Neckers, J. Pena, R. H. Jones, J. B. Field, *Biochem. Soc. Trans.* 752–754 (1990).

98. J. Kopecek, *J. Bioactive Compat. Polymers* 3: 17–26 (1988).

99. H. Brondsted, J. Kopecek, *Proc. Int. Symp. Control. Rel. Bioact. Mater.* 17: 128–129 (1990).

100. J. F. Bridges, J. F. Woodley, R. Duncan, J. Kopecek, *Intl. J. Pharm.* (in press).

101. J. D. Davies, E. Touitou, A. Rubinstein, European Patent Application No. 86309305.0 (1986).

102. E. Touitou, A. Rubinstein, *Int. J. Pharm.* 30: 95–99 (1986).

103. P. Gruber, M. A. Longer, J. R. Robinson, *Adv. Drug Del. Rev.* 1: 1–18 (1987).

9

Influence of Disease on Colonic Drug Absorption

Stuart A. Riley

Northern General Hospital, Sheffield, England

I. GENERAL CONSIDERATIONS

To the clinician, colonic drug delivery has several advantages over conventional oral dosing. First, transit through the colon is slow in comparison with other regions of the gastrointestinal tract. Slow-release formulations may, therefore, extend the duration of action of a rapidly absorbed drug and thereby facilitate once-daily dosing. Second, oral colonic delivery formulations may protect the upper gastrointestinal mucosa from the damaging effects of the drugs they carry. Third, the contents of the colon are less toxic to biologically active peptides than are the contents of the stomach and small intestine. By protecting such peptides from acid- and enzyme-related injury, colonic formulations make oral peptide dosing feasible. Finally, such formulations lead to much higher intra-colonic drug concentrations than do standard oral treatments. This is of particular benefit in the management of colonic diseases such as colitis, which respond best to high-dose topical treatment.

177

These features give colonic drug delivery a considerable potential in clinical medicine and its use will probably become more widespread. It is, therefore, important to be aware of potential interactions between disease and colonic drug absorption. Since these interactions are best considered in terms of altered physiology, it is first useful to recall the key determinants of colonic absorption that have been detailed in Chapter 1.

To be absorbed from the colon, drugs must first reach the colonic lumen. Two routes of access are available. The anal route is direct and reliable and minimizes the problems of upper gastrointestinal toxicity. However, drug delivery is limited to the rectum and distal colon and many patients find this route unacceptable for long-term dosing.

Oral colonic formulations are more acceptable to the patient and permit drug delivery to the whole of the colon. However, such formulations have to traverse the upper gastrointestinal tract and thereafter release their contents into the colonic lumen. The mechanisms of drug release within the colon depend on the type of formulation administered. Those currently available usually rely on osmotic effects, regional differences in luminal pH, or bacterial azocleavage. Once liberated within the colonic lumen, drugs have to cross the colonic mucosal barrier to gain access to the portal and systemic circulations. Mechanisms of mucosal permeation are complex and depend not only on the physicochemical characteristics of the drug but also on a wide range of physiological variables such as luminal availability, contact time with the colonic mucosa, bulk phase and microclimate pH, solvent drag effects, mucosal permeability and metabolism, and regional blood flow.

Surprisingly few studies have assessed the effect of disease on colonic drug absorption directly. An understanding of the physiological determinants of colonic drug absorption and an appreciation of disease-related changes in colonic physiology, however, may help predict when interactions are likely to occur.

II. DISEASE EFFECTS ON COLONIC DRUG DELIVERY

To reach the colon, orally administered formulations have to traverse the esophagus, stomach, and small intestine. Diseases of the upper gastrointestinal tract may therefore influence the delivery of drugs into the colonic lumen.

A. Esophageal Transit

Although diseases of the esophagus have no appreciable effect on drug delivery, slow-release and delayed-release formulations may lodge in the esophagus and cause local damage (1). Even in normal subjects the transit of large dosage forms may be greatly prolonged (more than 20 min) yet this causes surprisingly few symptoms (2). Patients with reflux esophagitis, esophageal stricture, achalasia, systemic sclerosis, and other motility disorders are particularly susceptible (3). It is important to instruct all patients to ingest large formulations while in the upright position and with at least 100 ml of water (4).

B. Gastric Emptying

The gastric emptying of a drug formulation is largely determined by the presence or absence of food in the stomach and the type of formulation ingested (5).

In both fasting and fed states the stomach selectively empties liquids while retaining solids. In the fasting state, liquids and small solids (less than 2 mm diameter) empty rapidly from the stomach. Larger solids, which include most of the solid oral colonic delivery formulations, are only emptied by the passage of migratory motor complexes, which sweep the contents of the stomach to the pylorus and deliver them into the duodenum about every 2 hr (6). In the fed state liquids and small solids (less than 2 mm diameter) are still emptied rapidly. Larger solids, however (particularly those over 7 mm in diameter), are mostly retained in the stomach until after the meal has emptied and the interdigestive migratory motor complexes restart (7). Since the gastric emptying of a meal may take several hours, the emptying of colonic formulations ingested with a meal may be considerably delayed. Furthermore, such formulations ingested with breakfast may be retained within the stomach by the regular intake of snacks and meals throughout the day and may not empty until later the same night (8).

A wide range of drugs (Table 1) and diseases (Table 2) may perturb gastric emptying and have the potential to influence colonic drug delivery. Rapid gastric emptying, however, is unlikely to have a major clinical effect on colonic drug delivery unless transit in other parts of the gastrointestinal tract is also rapid. Drugs and

Table 1 Drugs That Influence Gastric Emptying

Slow emptying	Speed emptying
Opiates	Metoclopramide
Anticholinergic drugs	Domperidone
Tricyclic antidepressants	Cisapride
Adrenergic agonists	Anticholinesterases
L-Dopa	Reserpine

diseases that slow gastric emptying, on the other hand, may greatly delay the onset of drug release (9). Furthermore, successive doses may accumulate in the stomach. If these then empty together, potentially toxic concentrations of drug may enter the systemic circulation. In rare cases, prolonged retention of many doses may lead to the formation of pharmacobezoars (10).

C. Small Intestinal Transit

Unlike the stomach, the small intestine does not differentiate between solids and liquids with respect to transit, and passage from the duodenum to the cecum usually takes 3–5 hr (11).

Few diseases speed small intestinal transit appreciably and rapid transit is unlikely to have a major effect on colonic drug

Table 2 Diseases Associated with Altered Gastric Emptying

Slow emptying	Rapid emptying
Mechanical obstruction	Duodenal ulcer disease
Gastric carcinoma	Zollinger-Ellison syndrome
Pyloric ulcer	Dumping syndrome
Gastroesophageal reflux	
Gastric ulcer and gastritis	
Diabetes and hypothyroidism	
Postgastric surgery	
Idiopathic and secondary	
pseudo-obstruction	
Malabsorption	

delivery unless transit in the colon is also rapid. Disorders that slow small intestinal transit time, however, such as scleroderma and intestinal pseudo-obstruction, have the potential to delay the onset of drug release from colonic formulations. Furthermore, diseases associated with small intestinal strictures, such as Crohn's disease, may predispose to impaction of solid formulations at areas of narrowing, with subsequent intestinal obstruction. Moreover, small intestinal ulceration or perforation may result when formulations lodge and release concentrated drug at one site over a prolonged time (12).

III. DISEASE EFFECTS ON COLONIC DRUG RELEASE

Once a colonic drug formulation has reached the cecum, it must release its contents into the colonic lumen. Factors that determine the onset and rate of drug release vary depending on formulation type.

A. Colonic Transit

Colonic transit is almost always important irrespective of formulation type (see section IIIC) allowing either more or less time for drug release in patients with diarrhea or constipation. Diarrhea (Tables 3 and 4) may be particularly troublesome and certain for-

Table 3 Drugs That Cause Constipation and Diarrhea

Constipation	Diarrhea
Opiates	Laxatives
Antacids (containing aluminium and calcium)	Alcohol
	Magnesium salts (e.g. antacids)
Anticholinergic drugs	Diuretics
Tricyclic antidepressants	Nonsteroidal anti-inflammatory drugs
Antiparkinsonian drugs	
Diuretics	Methyl dopa
Ganglion-blocking drugs	Digoxin
Anticonvulsants	Beta-blockers
Iron preparations	Theophylline
Monoamine oxidase inhibitors	Cimetidine
Anesthetic agents	Colchicine
	Olsalazine

Table 4 Diseases That May Cause or Be Complicated by Diarrhea

Acute diarrhea	Chronic diarrhea
Infections Viral	Inflammatory bowel disease
Bacterial	Irritable bowel syndrome
Parasitic	Colonic carcinoma
Travellers' diarrhea	Malabsorption
Food poisoning	Vagotomy or resection
	Thyrotoxicosis

mulations may pass intact through the length of the gastrointestinal tract (13).

B. Bulk Phase pH

Some drug formulations aim to deliver their contents into the colonic lumen by using pH-sensitive coat dissolution. Recent studies using free-floating, pH-sensitive radiotelemetry capsules show striking regional differences in bulk phase pH along the length of the gastrointestinal tract (Fig. 1). In healthy subjects, gastric pH is highly acidic (pH 1–2.5) but rises rapidly in the duodenum to 6.5 in the proximal jejunum and then gradually to 7.5 in the terminal ileum. Values fall to 6.5 on passing into the cecum and then rise again slowly passing from the right to left colon, with a final stool pH of approximately 7.0 (14). Important disease-related changes in upper gastrointestinal pH most often relate to gastric hypochlorhydria or achlorhydria. Atrophic gastritis, which is common in the elderly, and treatment with H_2 antagonists and proton pump inhibitors are the usual causes. Standard doses of H_2 antagonists, which cause only a modest rise in intragastric pH, appear to have little effect on drug release from pH-dependent mesalazine formulations (15). In patients with atrophic gastritis and in those receiving proton pump inhibitors, however, intragastric pH may be near neutral and there is probably a considerable risk of premature drug release from such formulations, particularly when they are retained within the stomach for any length of time. Further studies are needed to examine this possibility.

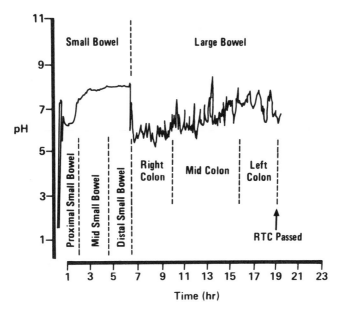

Figure 1 Luminal pH profile along the gastrointestinal tract of a healthy volunteer measured by a radiotelemetry capsule (RTC). (From reference 14).

C. Bacterial Actions

Some drug formulations rely on azo-linkage of the active drug to a carrier molecule to ensure colonic delivery. When taken by mouth, such compounds pass unchanged and largely unabsorbed through the stomach and small intestine. On reaching the cecum, bacterial azo-reductase enzymes cleave the azo bond, liberating the active drug from the carrier molecule (16). Sulfasalazine (5-aminosalicylic acid azo-linked to sulfapyridine) was the first azo drug to be developed and has been widely used for many years in the treatment of patients with inflammatory bowel disease.

Although azo-cleavage is usually highly efficient in healthy volunteers and in patients with quiescent inflammatory bowel disease, concomitant administration of antibiotics and diseases and drugs that accelerate colonic transit may greatly impair drug release. Houston et al. (17), for example, found that the administration of ampicillin to healthy volunteers resulted in an approximately 30% reduction in azo cleavage of sulfasalazine, and Van Hees et al. (18) reported a 50% reduction in azo-cleavage when intestinal transit was accelerated with bisacodyl. Furthermore, in patients with ulcerative colitis maintained on sulfasalazine, plasma sulfapyridine concentrations tend to fall during disease relapse, suggesting reduced azo-cleavage (19). This effect is probably related to the development of diarrhea rather than to the activity of the colitis per se.

Certain diseases predispose to bacterial colonization in parts of the gastrointestinal tract that are normally either sterile or have very low bacterial counts. Patients with gastric achlorhydria or hypochlorhydria may develop bacterial colonization in the stomach. Patients with small intestinal stasis because of strictures, surgically created blind loops, or hypomotility may likewise develop bacterial overgrowth of the small intestine (20). In both instances, the presence of bacteria may cause premature release of azo-linked drugs destined for colonic delivery.

IV. DISEASE EFFECTS ON COLONIC MUCOSAL PERMEATION

A. Colonic Transit

As noted previously, colonic transit may have an important effect on both colonic delivery and colonic drug release. It is also a key determinant of the extent of colonic drug absorption, since it governs the time available for drug contact with the colonic mucosa.

It is important to appreciate the highly variable nature of colonic transit. Even in healthy subjects, propulsion through the colon is characterized by short bursts of activity and longer periods of relative stasis. Total colonic transit is usually 20–30 hr, but may range from less than 1 hr to more than 60 hr. Disease-related changes in colonic transit, manifesting clinically as constipation or diarrhea, have a considerable potential to influence absorption.

Diarrheal illnesses reduce the time available for drug absorption and, therefore, decrease both the rate and extent of drug absorbed. Constipation, on the other hand, may increase the time available for absorption and predispose to drug accumulation. This is particularly likely when drug release is prolonged, since colonic transit time then becomes the rate-limiting factor to absorption. To avoid this problem, it has been suggested that the duration of drug release from colonic formulations should be limited.

Diarrhea and constipation are very common complaints. Population surveys show that up to 5% of subjects have loose or watery stools on more than one-quarter of occasions and up to 2% have more than three stools per day. Although colonic transit times have not been measured in such patients, they are probably rapid. Constipation, depending on definition, is likewise reported by 2–5% of healthy individuals (21, 22). Although the complaint of constipation is more prevalent in the elderly, in the absence of disease, bowel frequency is little different and transit times appear unrelated to age.

Diarrhea and constipation may be caused by a wide range of different drugs (Table 3) and diseases (Tables 4, 5), all of which have the potential to influence colonic drug absorption. It is important to realize, however, that the symptoms of diarrhea and constipation do not necessarily equate with changes in colonic transit. Studies of patients with nonorganic constipation show an inconsistent relationship between symptoms and colonic transit times. A third of such patients have delayed transit through the whole colon,

Table 5 Diseases That May Cause or be Complicated by Constipation

Colonic disorders	Anorectal disorders	Endocrine and metabolic	Neuropsychiatric
Simple constipation	Anal fissure	Pregnancy	Depression
Irritable bowel	Anal stenosis	Diabetes	Cerebrovascular
syndrome	Perianal	mellitus	disease
Idiopathic slow-	abscess	Hypothyroidism	Multiple sclerosis
transit	Descending	Hypercalcemia	Autonomic
constipation	perineum	Hypokalemia	neuropathy
Outlet obstruction	syndrome		Hirschsprung's
Intestinal pseudo-			disease
obstruction			Paraplegia
Colonic obstruction			
Strictures			
inflammatory			
neoplastic			
Volvulus			

a further third have delay in the distal colon and the remainder have normal transit time (23). Unfortunately, it is not possible to differentiate these groups on clinical grounds alone. Furthermore, in patients with irritable bowel syndrome, diarrhea and constipation are common symptoms, yet transit times often fall within the normal range. This probably reflects the wide range of normal and the subjective nature of symptom perception (24).

B. Luminal Metabolism

Although drug metabolism by colonic bacteria (see Chapter 2) has little influence on the absorption of drugs well absorbed in the proximal small intestine, it may influence drugs that are incompletely absorbed and those that undergo significant enterohepatic circulation. It probably has a major effect on the absorption of drugs delivered directly into the colon.

In health, inter- and intrasubject variability in colonic microflora is considerable and is influenced by factors such as age, dietary intake, ethnic group, and changes in colonic transit time and defecation frequency (25). Diseases that cause diarrhea (Table 4) and constipation (Table 5) probably have significant effects on colonic bacterial metabolism.

The effects of diarrhea and rapid colonic transit on intracolonic drug metabolism have not been studied. Any changes, however, are probably overshadowed by the effect of rapid transit on drug delivery, release, and mucosal permeation. Constipation and slowed colonic transit, on the other hand, may have significant effects on drug metabolism.

Many drugs are metabolized within the colon to less active or inactive products (see Chapter 2). Constipation may facilitate this process and less active drug may reach the distal colon. This may be particularly disadvantageous when a topical drug effect is required, as in patients with distal ulcerative colitis.

5-Aminosalicylic acid, one of the standard treatments for colitis, is metabolized by the colonic microflora and by the colonic mucosa to N-acetyl-5-aminosalicyclic acid (5-ASA), an inactive metabolite (26, 27). Recent studies (Fig. 2) suggest that slow colonic transit facilitates the acetylation process (28). Patients with constipation, common in a subgroup of patients with distal colitis, may therefore be at a therapeutic disadvantage since they inactivate more 5-ASA

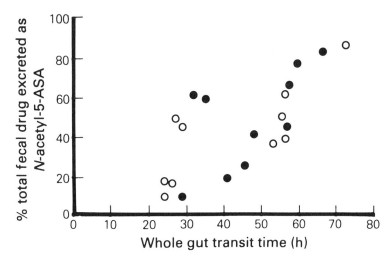

Figure 2   Relationship between fecal N-acetyl-5-aminosalicylic acid excretion and whole gut transit time in 10 patients with quiescent colitis maintained on a high- (●) and low-fibre (○) diet. (From reference 28).

before it reaches its site of action. Other drugs designed for local action in the distal colon may be similarly affected.

Some drugs are metabolized within the colonic lumen to toxic substances. Such products have been implicated in the side effects of L-dopa phenacetin, and chloramphenicol (56). Constipation may potentiate such toxicity.

Constipation may also facilitate enterohepatic circulation and thereby prolong the action of certain drugs. Drugs excreted in the bile as inactive glucuronate or sulfate conjugates are particularly susceptible to this effect since these may be deconjugated by bacteria on re-entering the colon, thereby liberating active drug for reabsorption.

Antibiotic exposure may also influence intracolonic drug metabolism. It is well recognized, for example, that antibiotic treatment may impair digoxin metabolism (29). Digoxin is inactivated in the colon by bacterial reduction of its lactone ring. Exposure to either erythromycin or tetracycline impairs this process and increased systemic absorption may result in toxicity.

Antibiotic treatment may also impair drug absorption through changes in enterohepatic drug circulation and the failure of certain oral contraceptive formulations, when administered with antibiotics, may relate to reduced bacterial hydrolysis of the steroid conjugates that usually undergo enterohepatic recycling. Other drugs that undergo recycling may be similarly affected.

Finally, it should be noted that diarrhea is not an infrequent complication of antibiotic therapy and this may impair colonic drug metabolism (see Chapter 10). Antibiotic-related diarrhea may result from a direct stimulatory effect of the antibiotic on gastrointestinal musculature, impairment of carbohydrate digestion, or the development of *Clostridium difficile* infection and pseudomembranous enterocolitis (30).

C. Bulk Phase and Microclimate pH

The importance of luminal pH on colonic drug release has been noted earlier. Bulk phase and microclimate pH are also important determinants of colonic mucosal drug permeation.

Most currently available pharmacologic agents are weak electrolytes and therefore exist as an equilibrium of ionized and nonionized forms. Since nonionized drugs permeate gastrointestinal epithelium more readily than their ionized counterparts, ambient pH may be an important determinant of absorption.

Both drugs and disease states may lead to changes in intracolonic and stool pH. Dietary fiber supplements, for example, may cause acidification of the intracolonic lumen and lower stool pH (32). Such supplements are commonly prescribed for the treatment of constipation and are taken by the population at large as a general health measure. Lactulose, also used in the treatment of constipation, causes a similar acidification of the intracolonic lumen and stool (33).

Many diarrheal illnesses lead to malabsorption of dietary carbohydrates. Subsequent bacterial fermentation of carbohydrates within the colon yields short-chain fatty acids, which also acidify the luminal contents and stool (34). Lactase deficiency, highly prevalent in many racial and ethnic groups, has a similar effect (35).

Although bulk-phase pH is an important determinant of luminal ionization, the concentration of nonionized drug in the juxtamucosal layer is determined by microclimate pH. Direct measurement of microclimate pH in human rectal biopsies shows values of pH 6.5, approximately 0.5 units below bulk phase (stool) pH (31). Little is known of the effects of disease on microclimate pH; further study would be of value.

D. Mucosal Permeability

Current theories describe the diffusion of drugs across epithelial barriers in terms of two parallel pathways of permeation. One, the transcellular pathway, is readily permeable to lipid-soluble drugs; the other, the paracellular pathway, permits the passage of small ionized drug forms. Little is known of the colonic transcellular permeability barrier in humans, but studies of transepithelial electrical resistance and paracellular probe permeation show clear regional differences in paracellular permeability along the length of the gastrointestinal tract, the colon exhibiting the highest paracellular resistance (36, 37). Furthermore, recent evidence suggests that paracellular permeability, at least in the small intestine, is not static but may be under physiological control. Disease states have the potential to perturb either pathway of permeation and may therefore, influence the absorption of both lipid-soluble and ionized species.

The movement of drugs across the colonic epithelium is dependent not only on the permeability properties of the epithelium but also on the surface area available for absorption. It is, therefore, important to remember that in patients who have undergone colonic resections, usually for the treatment of colonic carcinoma, complicated diverticular disease, or Crohn's disease, colonic drug absorption may be impaired. Following small bowel resection or bypass, the absorption of digoxin, phenytoin, hydrochlorthiazide, and penicillin is reduced (38–41). Similar effects are likely when colonic drug formulations are administered to patients who have undergone extensive colonic resection.

The permeability properties of the colonic mucosa may be changed by one of the many different forms of colitis (Table 6). Paracellular permeability appears particularly susceptible. In active colitis, the electrical resistance of the mucosa (a marker of paracellular permeability) falls and the permeation of probes such as chromium[51]-labeled EDTA increases (42). Such changes are probably mediated via cytokine secretion.

Changes in transcellular permeability in patients with active colitis have not been well studied. In severe disease, however, mem-

Table 6 A Classification of Colitis

Infections	Drug-induced
Viral	Penicillamine
Cytometalovirus	Gold
Herpes	Nonsteroidal anti-inflammatory agents
Bacterial	Methyl dopa
Salmonella	5-Aminosalicylic acid
Shigella	Antibiotics/*Clostridium difficile*
Yersinia	Ulcerative colitis and proctitis
E. coli	Microscopic colitis
Vibrio	Colonic Crohn's disease
Neisseria	Ischemic colitis and vasculitis
Chlamydia	Radiation colitis
Protozoan	Bechet's syndrome
Amoebiasis	Graft-versus-host disease
Schistosomiasis	Eosinophilic gastroenteritis

brane lipid composition may be changed and this may alter permeability. Furthermore, it has been suggested that subtle abnormalities of membrane permeability, as judged by chromium efflux from cells, may be present even in patients with quiescent colitis (43).

Although these experimental data suggest that colonic drug absorption should increase in patients with active colitis, clinical evidence suggests that this is not the case. Studies of the colonic absorption of 5-aminosalicylic acid (44) and prednisolone (45) enemas in patients with active and quiescent colitis show that absorption may be decreased in patients with active disease. Furthermore, the absorption of neomycin, when given by enema, is not significantly increased in patients with active ulcerative colitis (46). It seems likely, then, that factors other than mucosal permeability are important in determining colonic drug absorption in patients with active colitis. The colonic absorption of oral and rectal 5-aminosalicylic acid in patients with quiescent colitis appears very similar to that in healthy volunteers.

Recent studies have shown that chronic nonsteroidal anti-inflammatory drug treatment increases paracellular permeability in the small intestine and the colon (47) (see Chapter 10). The clinical effect of such changes on colonic drug absorption is unknown, although animal data suggest that macromolecular absorption may be increased. Patients receiving cytotoxic therapy may likewise have altered mucosal permeability and this has the potential to alter drug absorption (48).

E. Solvent Drag

Studies in rats show that transmucosal water flux may influence the absorption of drugs across the small intestinal mucosa (49, 50). Similar effects are likely in the colon.

In health, about 8 L of fluid enter the duodenum each day. This results from a combination of ingested fluids and salivary, gastric, pancreatic, biliary, and intestinal secretions. Most of this fluid is absorbed in the small intestine; 1.5 L enter the colon each day. Of this, approximately 90% is absorbed by the colonic mucosa, mostly in the proximal colon, and only 100–150 ml of fluid is passed in the stools each day.

In patients with small intestinal mucosal disease, absorption may be impaired and large volumes of fluid may enter the proximal colon. Although this may dilute intraluminal drug concentrations if the colon is healthy, much of this fluid will be absorbed and this may promote absorption (51). If, on the other hand, the colon is diseased, the mucosa may not absorb even normal amounts of fluid. Moreover, in some patients with active colitis, for example, the mucosa may secrete fluid into the colonic lumen (52). Luminal drug concentrations will be reduced by such secretion and drug absorption may be further impaired by the transmucosal movement of fluid from the mucosa into the colonic lumen. It is important to appreciate, however, that changes in solvent drag forces are unlikely to occur in isolation. In most patients with mucosal disease, changes in transit and luminal pH are also likely and the overall effect on colonic drug absorption will be determined by the interplay of all prevailing factors.

F. Mucosal Metabolism

The gastrointestinal mucosa is an important site of presystemic drug metabolism (see Chapter 2). Although mucosal metabolic activity is generally greater in the duodenum and proximal jejunum than in the distal ileum and colon, certain mucosal enzymes have high activity in colonic mucosa and have the potential to influence colonic drug absorption (53).

Data on the influence of disease processes on mucosal drug metabolism are scanty. Starvation and iron deficiency are thought to decrease small intestinal aryl hydrocarbon hydroxylase activity, and reduced conjugation of ethinyl estradiol has been reported in patients with celiac disease. There is also a growing body of data documenting the modification of small intestinal mucosal metabolism of one drug by the concomitant administration of another (53).

To date, only one study has assessed the effect of disease on colonic mucosal drug metabolism. Allgayer et al. investigated the ability of colonic mucosal biopsy homogenates, taken at colonoscopy, to acetylate 5-aminosalicylic acid (26). No important differences were found between biopsies taken from patients with mild and moderately active colitis and Crohn's disease and those taken from controls. The effect of active inflammatory bowel disease on the mucosal metabolism of other drugs is unknown.

G. Blood Flow

The mucosa along the length of the gastrointestinal tract is served by an abundant blood supply and this facilitates the rapid removal of absorbed compounds by maintaining a positive lumen-to-blood concentration gradient. In health, this gradient is so large that drugs that permeate the epithelial barrier slowly are unlikely to be influenced by changes in blood flow. The absorption of drugs that permeate the epithelium rapidly, however, may be blood-flow-dependent.

Knowledge of the human colonic mucosal blood flow is limited. Studies using endoscopic laser Doppler flowmetry suggest that mucosal blood flow is increased in patients with active ulcerative colitis, but not in Crohn's colitis (54), and is reduced by cigarette smoking (55). The effect of such changes on colonic drug absorption is unknown.

V. CONCLUSIONS

Colonic drug delivery has several advantages over conventional oral dosing and its use will probably become more widespread. It is therefore important for clinicians to be aware of the effects of drugs and disease states on colonic drug absorption. Unfortunately, few studies have addressed the issues directly. Insights into the key physiological determinants of absorption suggest that diseases that change gastrointestinal transit and luminal pH are most likely to influence colonic drug delivery, release, and absorption. However, many diseases have a diverse effect on gastrointestinal physiology and their influence on drug absorption is difficult to predict. Systematic studies of the relationships between disease-related changes in physiology and colonic drug absorption are urgently needed to improve our understanding of these important interactions.

REFERENCES

1. J. Pemberton, *Br. Heart J.* 32: 267 (1970).
2. K. T. Evans, G. M. Roberts, *Lancet* 2: 1237 (1976).
3. G. Vantrappen, J. Janssens, in *An Illustrated Guide to Gastrointestinal Motility* (D. Kumar, S. Gustavsson, eds.) John Wiley & Sons, Chichester (1988) p 313.
4. H. Hey, F. Jorgensen, K. Sorensen, H. Hasselbach, T. Warnberg, *Br. Med. J.* 285: 1717 (1982).
5. H. Minami, R. W. McCallum, *Gastroenterology* 86: 1592 (1984).
6. C. F. Code, J. A. Marlett, *J. Physiol. (Lond.)* 246: 289 (1975).
7. S. Davis, J. G. Hardy, A. Stockwell, M. J. Taylor, D. R. Whalley, C. G. Wilson, *Int. J. Pharm.* 21: 331 (1984).
8. E. Klaus, A. G. Press, S. Bollen, I. Schuhn, *Dig. Dis. Sci.* 36: 146 (1991).
9. W. S. Nimmo, *Clin. Pharmacokin.* 1: 189 (1976).
10. S. B. Raffin, in *Gastrointestinal Disease. Pathophysiology, Diagnosis, Management* (M. H. Sleisenger, J. S. Fordtran, eds.) W. B. Saunders, Philadelphia (1983) p 608.
11. S. Davis, J. G. Hardy, J. W. Fara, *Gut* 27: 886 (1986).
12. J. L. Shaffer, C. Higham, L. A. Turnberg, *Lancet* 2: 487 (1980).
13. K. W. Schroeder, W. J. Tremaine, D. M. Ilstrup, *N. Engl. J. Med.* 317: 1625 (1987).
14. G. Pye, D. F. Evans, S. Ledingham, J. D. Hardcastle, *Gut* 31: 1355 (1990).
15. E. H. Wiltink, C. J. Mulder, L. M. L. Stolk, R. Rietbroek, C. Verbeek, G. N. J. Tytgat, *Scand. J. Gastroenterol.* 25: 579 (1990).
16. S. A. Riley, L. A. Turnberg, *Q. J. Med.* 278: 551 (1990).
17. J. B. Houston, J. Day, J. Walker, *Pharmacology* 14: 395 (1982).

18. P. A. M. Van Hees, J. H. M. Tuinte, J. M. Van Rossum, J. H. M. Van Tongeren, *Gut* 20: 300 (1979).
19. A. K. Azad Khan, S. Truelove, *Gut* 21: 706 (1980).
20. G. L. Simon, S. L. Gorbach, *Gastroenterology* 86: 174 (1984).
21. D. A. Drossman, R. S. Sandler, D. C. McKee, A. J. Lovitz, *Gastroenterology* 83: 529 (1982).
22. A. M. Connell, C. Hilton, G. Irvine, J. E. Lennard-Jones, J. Misiewicz, *Br. Med. J.* 2: 1095 (1965).
23. A. Wald, *Arch. Intern. Med.* 146: 1712 (1986).
24. J. G. Hardy, E. Wood, A. G. Clark, J. R. Reynolds, *Eur. J. Nucl. Med.* 11: 393 (1986).
25. A. N. Alam, J. R. Saha, J. F. Dobkin, J. Lindenbaum, *Gastroenterology* 95: 117 (1988).
26. H. Allgayer, N. O. Ahnfelt, W. Kruis, U. Klotz, K. Frank-Holmberg, H. N. A. Soderberg, G. Paumgartner, *Gastroenterology* 97: 38 (1989).
27. R. A. Van Hogezand, P. A. M. Van Hees, J. P. W. M. Van Gorp, H. J. Van Lier, H. J. Bakker, P. Wesseling, U. H. G. M. Van Haelst, J. H. M. Van Tongeren, *Aliment. Pharmacol. Ther.* 2: 33 (1988).
28. S. A. Riley, I. A. Tavares, P. M. Bishai, A. Bennett, V. Mani, *Br. J. Clin. Pharmacol.* 32: 248 (1991).
29. J. Lindenbaum, D. G. Rund, V. P. Butler, D. Tse Eng, J. R. Saha, *N. Engl. J. Med.* 305: 789 (1981).
30. S. C. Rao, C. A. Edwards, C. J. Austen, C. Bruce, N. W. Read, *Gastroenterology* 94: 928 (1988).
31. N. I. McNeil, K. L. E. Ling, J. Wager, *Gut* 28: 707 (1987).
32. D. F. Evans, J. Crompton, J. Pye, J. D. Hardcastle, *Gastroenterology* 94: A 118 (1988).
33. R. L. Brown, J. A. Gibson, G. E. Sladen, B. Hicks, A. M. Dawson, *Gut* 15: 999 (1974).
34. W. J. Ravich, T. M. Baylees, in *Clinical Gastroenterology* (M. H. Sleisenger, ed.) W. B. Saunders, Philadelphia (1983) p 335.
35. H. A. Buller, R. J. Grand, *Annu. Rev. Med.* 41: 141 (1990).
36. G. R. Davis, C. A. Santa Ana, S. G. Morawski, J. S. Fordtran, *Gastroenterology* 83: 844 (1982).
37. V. S. Chadwick, S. F. Phillips, A. F. Hoffman, *Gastroenterology* 73: 257 (1977).
38. W. D. Heizer, T. W. Smith, S. E. Goldfinger, *N. Engl. J. Med.* 285: 257 (1971).
39. M. C. Kennedy, D. N. Wade, *Br. J. Clin. Pharmacol.* 7: 515 (1979).
40. J. P. Kampmann, H. Klein, B. Lumholtz, J. E. Molholm Hansen, *Clin. Pharmacokinet.* 9: 168 (1984).
41. L. Backman, B. Beerman, M. Groschinsky Grind, D. Hallberg, *Clin. Pharmacokinet.* 4: 63 (1979).

42. A. J. Geller, K. M. Das, *Curr. Opin. Gastroenterol.* 6: 561 (1990).
43. P. R. Gibson, E. Van De Pol, P. J. Barratt, W. F. Doe, *Gut* 29: 516 (1988).
44. M. Campieri, G. A. Lanfranchi, S. Boschi, C. Brignola, G. Bazzocchi, P. Gionchetti, M. R. Minguzzi, A. Belluzzi, G. Labo, *Gut* 26: 400 (1985).
45. D. A. H. Lee, M. Taylor, V. H. T. James, G. Walker, *Gut* 21: 215 (1980).
46. K. G. Breen, R. E. Bryant, J. D. Levinson, S. Schenker, *Arch. Intern. Med.* 76: 211 (1972).
47. A. P. Jenkins, D. R. Trew, B. J. Crump, W. S. Nukajam, J. A. Foley, I. S. Menzies, B. Creamer, *Gut* 32: 66 (1991).
48. G. R. Siber, R. J. Mayer, M. J. Levin, *Cancer Res.* 40: 3430 (1980).
49. H. Ochsenfahrt, D. Winne, *Naunyn-Schmiedebergs Arch. Pharmacol.* 281: 175 (1974).
50. H. Ochsenfahrt, D. Winne, *Naunyn-Schmiedebergs Arch. Pharmacol.* 281: 197 (1974).
51. S. Phillips, *Gastroenterology* 63: 495 (1972).
52. E. Q. Archampong, J. Harris, C. G. Clark, *Gut* 13: 880 (1972).
53. K. F. Ilett, L. B. G. Tee, P. T. Reeves, R. F. Minchin, *Pharmacol. Ther.* 46: 67 (1990).
54. D. E. Loft, C. J. Shorrock, S. A. Riley, W. D. W. Rees, *Gut* 29: A 1491 (1988).
55. E. D. Srivastava, M. A. H. Russell, C. Feyerabend, J. Rhodes, *Gut* 31: 1021 (1990).
56. J. W. Fara, in *Novel Drug Delivery* (L. F. Prescott, W. S. Nimmo, eds.) John Wiley & Sons, Chichester (1989) p 103.

10
Luminal Side Effects of Drugs in the Colon

Christoph H. Gleiter

Ciba-Geigy, Tübingen, Germany

I. INTRODUCTION

There is little comprehensive information on drugs or medicinal agents with potential noxious effects on the lower gastrointestinal tract. Injury to the colonic mucosa may occur as result of direct drug effects, systemic exposure, ischemia, or because of altered motility or toxin elaboration from changed bowel flora. Noxious effects of drugs on the lower gastrointestinal tract are rare. However, many of the cases described present severe, life-threatening problems. Therefore, despite their low incidence, their clinical importance is obvious.

Adverse reactions in the lower gastrointestinal tract may be caused by drugs used to treat gastrointestinal or any other diseases. The colonic mucosa shows only a limited number of reaction patterns. Causative factors are frequently difficult to determine, since drug-induced injuries may mimic other gastrointestinal diseases (1, 2). Therefore, in many cases the underlying cause can only be

197

linked with the symptoms by exclusion. The exact mechanism of action of many damaging medicinal agents remains unclear.

In this chapter the current knowledge about drug-induced injuries of the colon will be summarized. Common and rare causes will be discussed. It will focus on the direct and indirect *luminal* adverse effects on the colonic mucosa. Such topical side effects will probably become more important with the growing use of slow-release formulations that deliver larger amounts of drugs to the lower gastrointestinal tract than conventional dosage forms.

II. MECHANISMS OF TOPICAL DRUG TOXICITY IN THE LOWER GASTROINTESTINAL TRACT

A. Mechanical Lesions

Nonabsorbable or slow-release drug formulations often pass virtually unchanged through most of the gastrointestinal tract. Often they appear unchanged in the feces. Single entities are well imbedded in intestinal contents. Therefore, the mucosa will hardly be damaged. There is a theoretical possibility that ingestion of large quantities of such drug formulations at one time may cause mechanical injuries to the mucosa.

B. pH-Dependent Damage

Most damage due to pH changes occur in the upper gastrointestinal tract and are well documented. Very little is known about topical damage in the lower gastrointestinal tract due to drug-induced alterations. It appears unlikely that the pH would change in the colon following oral drug administration.

C. Osmolarity

Hyperosmolar enteral feedings or drugs can cause necrotizing enterocolitis. It is both a frequent and serious neonatal disorder in high-risk preterm infants (3). Involvement of the terminal ileum and colon is most frequent (4, 5). Many liquid drug formulations for infants are hyperosmolar (6–9) and can enhance food hyperosmolarity (6). Often high osmolarity is not due to the drug itself, but to additives and stabilizing components (10, 11). In adults, hyperosmolar water-soluble contrast media may lead to various degrees of colonic inflammation.

D. Formulation-Related Effects

Recent studies in animals and clinical observations have shown that the drug formulation can be crucial for direct damage of the mucosa in the lower gastrointestinal tract. Kircher et al. reported that the sodium salts of nonsteroidal anti-inflammatory drugs (NSAIDs) are more irritating than the free acid in an animal model (12).

The oral osmotic therapeutic system (OROS) is designed to maintain effective plasma concentrations for an extended period of time and to reduce side effects due to peak concentrations or rapid increase of plasma concentrations (see Chapter 8). Compounds with a short half-life profit from this controlled-release system. The OROS uses as osmotic driving agents either saturated solutions of the drug or its sodium and potassium salts. It consists of a nondigestible tablet with a semipermeable cover membrane. Its surface has a single perforation of defined size for drug release. After the system is ingested, an increasing internal pressure builds up due to water influx and the drug is released. The indomethacin system (Osmosin) has been reported to cause local damage of the lower gastrointestinal tract (14). Potassium salts also have been linked to ulcers or perforations in the lower gastrointestinal tract (13, 14).

Another potential mechanism for mucosal damage is the adhesiveness of the hydrophilic coating of such controlled-release formulations as demonstrated in vitro (15). Usually the coating is removed by body fluids in the upper gastrointestinal tract. Depending on the transit time through the gastrointestinal tract, the devices can pass into the lower tract with the coating largely intact. They can stick to the mucosa and may cause mucosal damage by constantly releasing the drug to a limited surface area of the gut.

Suppositories deliver drugs to the rectal mucosa and can lead to circumscript local damage (16–18), depending on the active ingredient (Table 1).

Table 1 Drugs in Suppository Form Causing Proctitis

Drug	Reference
NSAIDs	
Acetylsalicylic acid	109
Indomethacin	16, 110–112
Phenylbutazone	113, 114
Diclofenac	115, 116
Pirprofen	117
Acetaminophen	17
Others	
Ergot alkaloids	102–105, 114
Glycerin	96
Dextropropoxyphene	17

III. EXPERIMENTAL MODELS

A. Animal Models

Because of the increasing use of sustained- and controlled-release dosage forms, sizable amounts of drugs are delivered to the lower gastrointestinal tract. Therefore, attention has focused on direct, drug-induced irritation. During the past years, there has been an increasing interest in the systematic evaluation of such topical irritations.

Conventional tests are hampered by methodologic problems. Animals were used, which received the test drug orally or intragastrically (19). They had to be killed at arbitrary time points. Other techniques used direct application of test drugs on explanted gastric mucosa of dogs (20). None of these methods provided simultaneous assessment of the damage produced by two or more agents applied to adjacent areas of the intestinal mucosa.

In recent years, advanced animal models have been developed to assess the potential for such injuries. For the upper gastrointestinal tract, cat esophagus (21) or stomach (22) models have been widely used to assess topical drug effects. Fara et al. adapted the Alphin-Droppleman cat gastric mucosa model for the rabbit colon (23). This in situ test allows exposure of adjacent mucosal sites to several drugs or various drug formulations at the same time. Colonic segments are opened along the antimesenteric side. The exposed mucosa is clamped in a multicell plexiglass perfusion chamber. Changes of the drug-exposed segments can be scored by direct macroscopic observation during the experiment (12, 24). The grading scales for mucosal damage are those proposed by Carlborg and Densert (21). The model appears to be sensitive, yields reproducible results, and can be used for comparison of different solid and liquid dosage forms at the same time. Disadvantages are that with this model the length of contact time with the human mucosa is probably overestimated since normal residence time and motility are not taken into account. Another drawback is the fact that the recovery of damaged bowel segments cannot be observed (23).

Colonoscopy in rats and in larger experimental animals is a method that allows direct inspection, biopsy of the mucosa, and a follow-up assessment of the lesions (25).

Measurement of intestinal fluid transport can be a probe for functional alterations of the intestinal mucosa cells by drugs or drug formulations causing diarrhea. Such studies have been performed with in situ or in vitro preparations of hamster intestine (26).

For antibiotic-associated colitis, a hamster model has been developed. Its relevance for the investigation of human disease has been proven by identifying the toxins of *Clostridium difficile* as the cause of this condition (27).

B. Models in Humans

There have been attempts to develop human models for the noninvasive evaluation of the mucosal irritation potential of drugs. The human buccal assay has been studied intensively as a test method for local lesions by immediate mucosal contact of drugs anywhere in the gastrointestinal tract. The test drugs are put into plastic cups. The cups are applied to the mucosa of the lower lip. The sensory innervation of the mucosa and the tactile sensitivity of the tongue allows continuous self-monitoring by the volunteers. Changes are often reported earlier by the subject than observed by the investigator. Frequent inspections provide a continuous recording of surface changes during both test and healing period (28, 29). However, buccal cells are different from most other mucosal cells in the gastrointestinal tract. This limits the predictive value of the test, particularly for the lower parts of the gastrointestinal tract.

In a recent comparative study, no correlation was shown between buccal lesions and mucosal damage of the upper gastrointestinal tract (30). There are no studies correlating alterations of buccal surface and lesions in the lower gastrointestinal tract.

The measurement of changes in buccal mucosal potential difference following mucosal lesions caused by certain drugs (e.g., NSAIDs) has never been validated with respect to its predictive value for colonic lesions (31). Taken together, at present no valid, easy-to-perform human model is available.

IV. TOPICAL SIDE EFFECTS IN THE LARGE INTESTINE

A. Antibiotics

Many antibiotic agents can cause indirect side effects in the lower gastrointestinal tract (32). Three different nosologic entities have been described: antibiotic-associated diarrhea, antibiotic-associated colitis, and hemorrhagic colitis. The most frequently implicated antibiotics are listed in Table 2.

1. Antibiotic-Associated Diarrhea

Orally administered antibiotics, particularly those with a broad spectrum, may cause diarrhea without signs of inflammation (Tab. 2). There are different theories as to the cause. After changes of the normal intestinal flora, superinfection with pathogenic organisms (predominantly *Clostridium difficile*) is held to be responsible (32, 33).

Another theory draws conclusions from analogies with animals raised in an abacterial environment. These animals have mild diarrhea during their entire lifetime. This is probably due to the lack of bacterial degradation of mucopolysaccharides and proteins. These compounds keep water in the intestinal lumen by osmosis and lead to diarrhea. Similar symptoms are observed in rats and mice fed antibiotics (34). A further explanation of this condition may be an altered pattern of bile acids in the feces leading to biliary diarrhea (35). The most recent theory suggests that antibiotic-associated diarrhea might be secondary to impaired colonic carbohydrate fermentation. Characteristics are reduced fecal concentrations and

Table 2 Antibiotics Frequently Causing Topical
Adverse Effects in the Colon

Drug	Reference
Antibiotic-associated diarrhea	
Cotrimoxazole	54
Penicillin	54
Ampicillin	54, 118, 119
Flucloxacillin	54
Cefuroxime	54
Lincomycin	118
Clindamycin	118, 119, 120
Antibiotic-associated colitis	
Ampicillin	119
Cephradine	121
Cephalosporin	122
Cotrimoxazol	123
Lincomycin	41
Clindamycin	42, 43, 124
Metronidazole	125
Chloramphenicol	37
Aureomycin	37
Ofloxacin	126, 127
Ciprofloxacin	128, 129
Segmental hemorrhagic colitis	
Penicillin	56
Amoxicillin	55, 56, 57, 130
Ampicillin	56

production rates of short-chain fatty acids and lactate (36). In general, symptoms are reversible after cessation of antibiotic treatment and normalization of intestinal flora (34).

2. Antibiotic-Associated Colitis

The more serious topical antibiotic adverse effect is antibiotic-associated colitis and its most severe form, pseudomembraneous colitis. This condition was recognized increasingly in the early 1950s, after the introduction of aureomycin and chloramphenicol (37). *Staphylococcus aureus* was considered to be responsible (38–40) until *Staphylococcus aureus* was demonstrated to increase without causing symptoms during almost any antibiotic therapy. Other antibiotics described as causing pseudomembranous colitis were lincomycin (41) and clindamycin (42). Half of the patients with diarrhea while taking clindamycin were shown by colonoscopy to have pseudomembranous lesions (43). Meanwhile, almost any kind of antibiotic has been linked with antibiotic-associated colitis (Table 2).

The history of the disease and the underlying cause have been clarified in recent years. The main work on the discovery of the responsible pathogen was done in an animal model (27, 44). Bartlett showed that the toxins of *Clostridium difficile* are the cause of antibiotic-associated colitis and the damage to the intestinal mucosa (45).

In newborns, *Clostridium difficile* belongs to the physiological intestinal flora (46). In adults, only 3% of the population carry small numbers of this strain (47). The induction of antibiotic-associated colitis is thought to be caused by the release of toxins from the bacterial wall. Apparently, it is not dependent on the number of bacteria, which usually is rather small (48). The bacterial wall is damaged by minimal concentrations of most antibiotics. This may explain the breakout of antibiotic-associated colitis days or weeks after cessation of oral or parenteral antibiotic therapy (48). These observations argue against older theories, which assumed a simple overgrowth of normal flora by pathologic bacteria (48).

Watery and voluminous diarrhea occurs in most patients. Fewer than 20% of patients experience bloody diarrhea (49). On endoscopic examination, the anatomical lesions range from minimal mucosal erythema to extensive pseudomembranous lesions (50, 51). Up to 30% of patients are thought to have right-sided colonic disease with rectal sparing (52). First relapses may occur in up to 20% of patients within days or weeks. Second relapses may occur in 20–40% of patients. Diagnosis has been improved by assays for the pathognomonic cytotoxin of *Clostridium difficile* (50). Treatment of antibiotic-associated colitis consists of elimination of the pathogenic agent and symptomatic treatment (49, 53). Responses to adequate therapy are usually prompt (53). The former 20% mortality has decreased considerably because of earlier diagnosis and therapy (50).

Cases with *Clostridium perfringens*-associated antibiotic-associated colitis have also been described (54).

3. Segmental Hemorrhagic Colitis

This characteristic form of colitis is thought to be associated with the use of ampicillin or other penicillin derivatives such as amoxicillin or dicloxacillin (55, 56) (Table 2). The clinical picture and outcome differ clearly from antibiotic-associated colitis. Therefore, it is believed to represent a separate nosologic entity. Patients experience abdominal cramps and bloody diarrhea within hours to a few days after the start of oral penicillin treatment. In contrast to antibiotic-associated colitis, *Clostridium difficile* or its toxin could not be isolated and the colon shows segmental mucosal hemorrhage, predominantly in the ascending part. Shortly after discontinuation of the penicillin derivative, the patients recover (34, 55–57). There are no markers for an allergic pathogenesis. The current hypothesis on the cause of this syndrome is that extensive antibiotic therapy causes ischemia by overgrowth of enterotoxin-producing bacteria or absorption of bacterial endotoxins from the colon. A similar clinical feature has been demonstrated recently after an outbreak of *E. coli*-associated hemorrhagic colitis. In these cases, verotoxin-producing *Escherichia coli* were isolated (58).

B. Antirheumatic Drugs

1. Nonsteroidal Anti-Inflammatory Drugs

In contrast to extensive investigations in the upper gastrointestinal tract, the mechanisms by which NSAIDs can damage the colonic mucosa have been less intensively studied. This may be due to the relative inaccessability of the large bowel. Nevertheless, adverse effects of NSAIDs in the lower intestinal tract are described in increasing numbers since fatal incidents involving slow-release NSAID have drawn attention to this problem (14). Epidemiologic studies in patients have demonstrated a link between NSAID intake and colonic lesions (59–62).

Animal experiments in various aspects have shown adverse effects of NSAIDs on the mucosa of the large intestine (19, 63). A likely mechanism of action may be by reduction of prostaglandin synthesis (63, 64).

Other mechanisms of action were demonstrated in animal experiments. NSAIDs, particularly fenamates, had effects on enterocytes and intestinal secretion similar to laxatives (26). In 12% of patients treated with mefenamic acid, diarrhea was the major gastrointestinal side effect (65).

Bjarnasson et al. (66) have shown that long-term NSAID treatment leads to a long-lasting inflammation of human intestine (up to 16 months) and enhances the permeability of the colon (67, 68).

Topical damage of the mucosa may also be due to high local concentrations of active compound delivered by slow-release formulations. In patients with diverticulitis, perforation of the colon has been described (69).

Other topical lesions, probably related to NSAIDs, are colonic strictures (70). A localized, submucosal, fibrous reaction following NSAID-mediated ulcers has been proposed as possible cause. It is thought that topical damage is a very likely explanation for the observed changes (70). A well-documented case had strictures in the cecum and ascending colon during treatment with slow-release diclofenac (70).

Direct contact with NSAID-containing suppositories or enemas is documented as possible cause of proctitis (Table 1). It is also conceivable that parent compounds or their active metabolites (e.g., indomethacin or piroxicam) with an appreciable enterohepatic circulation can reach the large intestine in amounts that can cause a direct damage of the mucosa (71). Indomethacin is eliminated to 100% as glucuronide in the bile (71). These glucuronides may be split in the colon and indomethacin reactivated in the lower gastrointestinal tract. Animal experiments show that indomethacin-induced intestinal lesions can be prevented by fasting or interruption of the enterohepatic circulation (72, 73). Indomethacin and piroxicam are known to cause colitis in humans (60, 74). The NSAID sulindac is a prodrug. It is converted to sulfide in the liver and seems to be the important anti-inflammatory metabolite. Sulindac sulfide is predominantly excreted with bile and subject to enterohepatic circulation. In the gastrointestinal lumen it exists mostly in the form of the inactive sulindac sulfoxide (65). This fact has been used to explain its better gastrointestinal tract tolerability than that of other NSAIDs undergoing enterohepatic circulation (65).

Sometimes medication with NSAIDs unmasks silent ulcerative colitis or Crohn's disease (75). Oral and rectal salazosulfapyridine (SASP) can cause deterioration of ulcerative colitis (76). The cleavage of the prodrug SASP into 5-aminosalicylic acid (5-ASA) and sulfapyridine occurs in the colon under the influence of the colonic flora. The inflammation seems to be caused by the NSAID-like 5-ASA and not by sulfapyridine (77).

2. Gold

For more than 50 years, gold has been used for treatment of chronic polyarthritis. The first preparations, gold sodium thimalate and aurothioglucose, could be given only parenterally. Much of the gold is retained in the body for long periods because of its slow excretion. Twice the amount of gold appears in urine than in the feces. Adverse reactions in the gastrointestinal tract occur only occasionally. Colitis is a rare but frequently fatal complication of chrysotherapy. Almost 20 cases of this adverse reaction have been reported in the medical literature (78–82). Severe watery and/or bloody diarrhea is frequent. Spontaneous perforation of the colon has been reported (83, 84). On macroscopic examination it is hard to differentiate from ulcerative colitis (85). The mortality rate approaches 30%. There has been a decrease in its incidence in recent years due to the physicians' greater awareness and an improved therapeutic regimen (82). Cromolyn sodium has been found useful in the therapy of gold-induced enterocolitis, suggesting an allergic cause (86, 87). Also, eosinophilia and the occurrence after low-dosage administration in many cases support the theory of allergic origin (85, 49). As an alternative, the colitis may be a direct toxic effect of elemental gold on the rapidly proliferating intestinal cells, as occurs with other heavy metals (83). Since 1972, the orally absorbed gold preparation auranofin (tri-ethylphosphin gold) has been available for the treatment of rheumatoid arthritis. A much lower percentage of gold is retained in the body with this drug than with the parenteral compounds. During long-term therapy, up to 88% is excreted in the feces and 12% in urine (88). Gastrointestinal side effects are the most common adverse reactions seen with auranofin. About 30–40% of all patients taking auranofin experience loose stools or diarrhea (88). These symptoms may be transient and respond to temporary discontinuation, reduction in dosage, or symptomatic treatment. Diarrhea secondary to auranofin has not been associated with weight loss, gastrointestinal (GI) bleeding, malabsorption, or histologic abnormalities (88). It has been suggested that auranofin causes its gastrointestinal adverse effects by a direct effect on sodium transport. Inhibition of this transport reduces the sodium-dependent transfer of monosaccharides and amino acids at the surface membrane of enterocytes (89). Animal experiments support this explanation (90).

There is also evidence from studies in animals and humans that auranofin increases intestinal permeability of the enterocyte membrane (89, 91). These changes are seen with auranofin but not with intramuscular gold preparations. Therefore, it is also conceivable that the carrier molecule and not the gold itself is responsible for the adverse reactions. However, the laxative effects of auranofin appear to be pathophysiological rather than idiosyncratic. Histologic examination of intestinal biopsies from patients receiving auranofin were normal, even during periods of diarrhea (89). It is not known yet which of the mechanisms accounts for the gastrointestinal symptoms caused by auranofin. In contrast to the laxative effects of auranofin, there are reports of cases with enterocolitis following oral gold therapy (81, 92, 93, 94). The pathophysiological and clinical features of these cases closely resemble those after parenteral gold therapy and are possibly of allergic nature. This is supported by one patient who responded favorably to treatment with cromolyn sodium (93).

Patients developing gastrointestinal symptoms during any kind of gold therapy need prompt attention and the type of adverse gastrointestinal reaction must be carefully evaluated.

C. Laxatives

1. Stimulant Laxatives

Most laxatives of interest for this review are of the stimulant type. These are diphenylmethane derivatives, such as bisacodyl or phenolphthalein and anthraquinone glycosides, such as senna. These laxatives are believed to be active only in the colon. They induce fluid and electrolyte accumulation in the small and large intestine of animals and humans.

Common mechanisms of action of these compounds are inhibition of active absorption by inhibition of enterocyte Na^+/K^+-ATPase, stimulation of active secretion of fluid and electrolytes by prostaglandin E biosynthesis, and increase of adenylate cyclase activity. In addition, it has been assumed that stimulant laxatives cause a change in mucosal permeability by facilitating permeability through the paracellular pathway, rendering the tight junctions more leaky (95).

Bisacodyl, a diphenylmethane derivative, has also been shown to stimulate the mucosal nerve plexus of the colon (96). It is not known whether this may also be a mechanism of action of anthraquinone derivatives. Phenolphthalein appears to interfere with glucose absorption and thus causes accumulation of intraluminal fluid (96). Chronic use of stimulatory laxatives may lead to the so-called cathartic colon. Clinical signs are worsening constipation, ill-defined lower abdominal pain, and bloating (49). In addition, pseudo-obstructions can be documented radiographically. This syndrome is thought to be due to loss of colonic myenteric neurons (49).

Another syndrome is characterized by diarrhea, weakness, hypokalemia, and protein-losing enteropathy (49). Experimental findings with bisacodyl in rats indicate that chronic administration reduces the desired effect on water and sodium secretion, whereas the enhanced potassium secretion remains unchanged (95). This explains the vicious circle of constantly decreasing effectiveness and increasing amounts of laxatives necessary to obtain bowel movements.

Pathologic absorption has been described as a side effect of chronic anthraquinone intake. It is thought to be caused by damage to the ultrastructure of absorptive cells of the colonic mucosa (97).

Melanosis coli is specific to long-term anthraquinone use. It is considered to be a medical curiosity without clinical sequelae or functional disturbances. The colon (mainly rectum and cecum) shows a brownish pigmentation. This results from macrophages in the lamina propria containing brown, lipofuscin-like pigment. Melanosis coli occurs within 1 year and disappears 6–11 months after cessation of treatment (49, 97).

Anthraquinones appear to damage the myenteric plexus during chronic use (97–99). Pigmentation and damage to the intrinsic innervation of the colon are believed to be the result of a common cytologic response involving lysosomes in macrophages and nervous tissue (98).

2. Lactulose

Lactulose is a synthetic disaccharide consisting of fructose and galactose and is used as a laxative. It is not broken down by disaccharidase enzymes in the small intestine and passes unchanged into the colon. There, it leads to an increase of stool osmolarity (100). This is thought to increase intraluminal pressure and to stimulate peristalsis. During its passage in the colon it is cleaved by bacteria (*Lactobacillus, Streptococcus faecalis*) to lactic acid and acetic acid. These organic acids exert an additional secretagog effect on colonic mucosa. Side effects of lactulose treatment are flatulence, cramps, abdominal distention with colics, diarrhea, and electrolyte imbalances (100).

3. Glycerin

Glycerin is a hyperosmotic laxative used only rectally. It is considered to be safe but may occasionally cause rectal irritation (96).

D. Enemas

Enema preparations have been reported to cause severe damage to the rectal or colonic mucosa.

Soap-containing enemas were very commonly used for many years. As many as 30% of patients receiving soap enemas experience immediate rectal irritation that may last for several days to months (85, 49). This represents a direct toxic or detergent action on the mucosa. The severity of the acute inflammatory reaction is related to the concentration of the enema and to the duration of contact (49). Alkaline soaps cause liquefaction necrosis, supporting extension of the necrotic process into deeper tissues (85, 101).

Barium and diatrizoate derivatives with osmolarities between 650 and 1900 mOsmol/L may cause various degrees of colitis up to necrotization with perforation proximal to an obstruction (85). Therefore, in the presence of obstruction, the use of water-soluble contrast media has been advocated.

E. Miscellaneous

1. Ergotamine

Some case reports provide evidence for anorectal ergotism as a cause of solitary rectal ulcers (102–105). These cases are most likely due to excessive use of suppositories containing ergot alkaloids. Most patients had had migraine headache and the amount of agent used exceeded by far the highest recommended dosage. Based on endoscopic and histologic findings, ergotamine-induced lesions cannot be differentiated from the solitary rectal ulcer syndrome (105). However, the clinical course appears to be different. In patients with solitary rectal ulcer syndrome, symptoms and endoscopic findings persist for years whereas ergot-induced ulcers heal within 8 weeks after discontinuation of the drug (105).

2. Bile Acids

Bile acids are used for the medical dissolution of cholesterol gallstones. One of their most prominent adverse effects is watery diarrhea. The secretory properties of these bile acids commonly found in the human intestine could be demonstrated in healthy volunteers by colonic perfusion studies. The dihydroxy bile acids, chenodeoxycholic and deoxycholic acids, are able to inhibit water and electrolyte absorption at concentrations of 1–3 mM and induce secretion at concentrations of 3–5 mM (106). The secretory activity may be due to topical cAMP stimulation (106). Diarrhea occurring after chenodeoxycholic acid has an incidence of 40–50% at dosages of 15 mg/kg body weight, causing net water secretion within the colon (107). Diarrhea may respond to a reduction in dosage (106).

More recently, ursodeoxycholic acid was investigated. This bile acid has in patients an incidence of diarrhea below 5% (108). Some groups have advocated a combination of chenodeoxycholic and ursodeoxycholic acid for dissolution of gallstones. This dual treatment appears to be more effective in the reduction of biliary saturation than one agent alone and causes only mild gastrointestinal side effects (108).

3. Antacids

Magnesium salts are widely used as antacids in various galenic forms. These salts may reach the colon. They produce diarrhea believed to be due to large amounts of osmotically active material.

Particularly magnesium trisilicate is given in larger amounts because of its low neutralizing capacity (33).

ACKNOWLEDGMENT

This chapter is dedicated to Dr. Melvin E. Gleiter, Emeritus Professor, University of Wisconsin, Eau Claire, Wisconsin, on the occasion of his 65th birthday.

REFERENCES

1. A. K. Banerje, *Br. Med. J.* 298: 1539 (1989).
2. R. Ottenjahn, *Dtsch. Med. Wochenschr.* 110: 1225 (1985).
3. S. L. Topalian, M. M. Ziegler, *J. Surg. Res.* 37: 320 (1984).
4. T. V. Santulli, J. N. Schullinger, W. C. Heird, *Pediatrics* 55: 376 (1975).
5. F. Hansbrough, C. J. Priebe, K. W. Faltermann, *Am. J. Surg.* 145: 169 (1983).
6. A. E. Mutz, M. W. Obladen, *Pediatrics* 75: 371 (1985).
7. M. Obladen, A. Mutz, *Monatschr. Kinderheilkd.* 133: 669 (1985).
8. K. C. White, K. L. Harkavy, *Am. J. Dis. Child.* 136: 931 (1982).
9. J. A. Ernst, J. M. Williams, M. R. Glick, J. A. Lemons, *Pediatrics* 72: 347 (1982).
10. A. M. Glasgow, R. L. Boeckx, M. K. Miller, M. G. MacDonald, G. P. August, *Pediatrics* 72: 353 (1983).
11. J. L. Brown, *N. Engl. J. Med.* 407: 439 (1983).
12. C. H. Kirchner, R. B. Smith, B. Gemzik, C. R. Overvold, *Toxicol. Pathol.* 15: 370 (1987).
13. M. D. Rawlins, in *Novel Drug Delivery and its Therapeutic Application* (L. F. Prescott, W. S. Nimmo, eds.) John Wiley and Sons, Chichester (1989) p 235.
14. J. L. Bem, R. D. Mann, R. Coulson, *Pharmaceut. Med.* 3: 35 (1988).
15. A. Noormohammadi, *Pharmaceut. J.* 232: 527 (1984).
16. N. Levy, E. Gaspar, *Lancet* 1: 577 (1975).
17. G. Cheymol, M. Biour, J. Jablonka, *Ann. Gastroenterol. Hepatol.* 22: 217 (1986).
18. B. Wörmann, W. Höchter, R. Ottenjahn, *Dtsch. Med. Wochenschr.* 110: 1504 (1985).
19. T. H. Kent, R. M. Cardelli, F. W. Stamler, *Am. J. Pathol.* 54: 237 (1969).
20. F. O. Stephens, G. W. Milton, F. Loewenthal, *Gut* 7: 223 (1966).

21. B. Carlborg, O. Densert, *Eur. Surg. Res.* 12: 270 (1980).

22. R. S. Alphin, D. A. Droppleman, *J. Pharm. Sci.* 60: 1314 (1971).

23. J. W. Fara, L. D. Anderson, A. G. T. Casper, R. E. Myrback, *Pharmaceut. Res.* 5: 165 (1988).

24. J. W. Fara, L. D. Anderson, A. G. T. Casper, R. E. Myrback, *Toxicologist* 6: 307 (1986).

25. N. S. Mann, L. M. Demers, *Gastrointest. Endoscop.* 29: 77 (1983).

26. G. W. Gullikson, M. Sender, P. Bass, *J. Pharmacol. Exp. Ther.* 220: 236 (1982).

27. A. B. Price, *Clin. Gastroenterol.* Suppl. 1: 151 (1980).

28. V. Place, P. Darley, K. Baricevic, A. Ramans, B. Pruitt, G. Guittard, *Clin. Pharmacol. Ther.* 43: 233 (1988).

29. J. A. Roth, P. Valdes-Dapena, P. Pieses, E. Buchmann, *Gastroenterology* 44: 146 (1963).

30. K. H. Antonin, P. R. Bieck, *Eur. J. Clin. Pharmacol.* 36: Suppl. A144 (1989).

31. G. J. Huston, *Br. J. Clin. Pharmacol.* 11: 528 (1981).

32. *Med. Lett* 31: 94 (1989).

33. D. N. Bateman, in *Textbook of Adverse Drug Reactions* (D. M. Davies, ed.) Oxford University Press, New York (1985) p 239.

34. K. Loeschke, *Klin. Wochenschr.* 58: 337 (1980).

35. A. F. Hoffmann, *J. Infect. Dis.* Suppl. 135: S126 (1977).

36. M. R. Clausen, H. Bonnen, M. Tvede, P. B. Mortensen, *Gastroenterology* 101: 1497 (1991).

37. L. Reiner, M. J. Schlesinger, G. M. Miller, *Arch. Pathol.* 54: 39 (1952).

38. W. H. Daring, A. H. Baggenstoss, L. A. Weed, *Gastroenterology* 38: 441 (1960).

39. M. Y. Khan, W. H. Hall, *Ann. Intern. Med.* 65: 1 (1966).

40. J. G. Bartlett, *Surv. Dig. Dis.* 1: 54 (1983).

41. M. B. Robertson, K. J. Breen, P. V. Desmond, M. L. Mashford, A. M. McHugh, *Med. J. Aust.* 1: 243 (1977).

42. F. J. Tedesco, *Dig. Dis.* 21: 26 (1976).

43. F. J. Tedesco, R. W. Barton, H. D. Alpers, *Ann. Intern. Med.* 81: 429 (1974).

44. T. W. Chang, J. G. Bartlett, S. L. Gorbach, A. B. Onderdonk, *Infect. Immun.* 20: 526 (1978).

45. J. G. Bartlett, T. W. Chang, S. L. Gurwith, S. L. Gorbach, A. B. Onderdonk, *N. Engl. J. Med.* 298: 531 (1978).

46. I. C. Hall, E. O. O'Toole, *Am. J. Dis. Child.* 49: 390 (1935).

47. W. L. George, R. D. Rolfe, M. E. Mulligan, S. M. Finegold, *J. Infect. Dis.* 140: 266 (1979).

48. R. Auckenthaler, *Schweiz. Rundschau. Med.* 71: 87 (1982).

49. J. H. Lewis, *Am. J. Gastroenterol.* 81: 819 (1986).

50. J. G. Bartlett, *DM* 30: 1 (1984).
51. R. L. Gebhard, D. N. Gerding, M. M. Olson, L. R. Peterson, C. J. McClain, H. J. Ansel, M. J. Shaw, M. L. Schwartz, *Am. J. Med.* 78: 45 (1985).
52. F. J. Tedesco, J. K. Corless, R. E. Brownstein, *Gastroenterology* 83: 1259 (1982).
53. J. G. Bartlett, *Gastroenterology* 89: 1192 (1985).
54. S. P. Boriello, H. E. Larson, A. R. Welch, R. F. Barclay, M. F. Stringer, B. A. Bartholomew, *Lancet* 1: 305 (1984).
55. M. Heer, H. Sulser, A. Hany, *Schweiz. Med. Wochenschr.* 119: 733 (1989).
56. R. B. Toffler, E. G. Pingoud, M. I. Burell, *Lancet* 2: 707 (1978).
57. C. Mrowka, R. Münch, M. Rezzonico, P. Greminger, *Dtsch. Med. Wochenschr.* 115: 1750 (1990).
58. A. O. Carter, A. A. Borzyk, J. A. K. Carlson, B. Harvey, J. C. Hockin, M. A. Karmali, C. Krishnan, D. A. Korn, H. Lior, *N. Engl. J. Med.* 317: 1496 (1987).
59. R. Elkaim, P. Brifford, J. C. Lapraz, F. Privat, M. Rometti, L. Euller, *Gaz. Med. Fr.* 87: 894 (1980).
60. M. J. S. Langman, L. Morgan, A. Worrall, *Br. Med. J.* 290: 347 (1985).
61. D. S. Rampton, *Scand. J. Gastroenterol.* 22: 1 (1987).
62. T. F. Ovokaitys, A. C. Caruso, J. N. Cooper, *Gastroenterology* 88: 1528 (1985).
63. A. Robert, *Gastroenterology* 69: 1045 (1974).
64. D. S. Rampton, T. B. Barton, *Agents Actions* 14: 715 (1984).
65. L. S. Simon, J. A. Mills, *N. Engl. J. Med.* 302: 1237 (1980).
66. I. Bjarnason, G. Zanelli, T. Smith, P. Prouse, P. Williams, P. Smethurst, G. Delacey, M. J. Gumpel, A. J. Levi, *Gastroenterology* 93: 480 (1987).
67. I. Bjarnason, P. Williams, A. So, G. D. Zanelli, A. J. Levi, J. M. Gumpel, T. J. Peters, B. Ansell, *Lancet* 2: 1171 (1984).
68. H. Yaginuma, T. Nakata, H. Toya, T. Murakami, M. Yamazaki, A. Kamada, *Chem. Pharm. Bull.* 29: 2974 (1981).
69. T. K. Day, *Br. Med. J.* 287: 1671 (1983).
70. T. Huber, C. Ruchti, F. Halter, *Gastroenterology* 100: 1119 (1991).
71. K. Brune, *Agents Actions* Suppl. 17: 59 (1985).
72. K. W. Sommerville, C. J. Hawkey, *Postgrad. Med. J.* 62: 23 (1986).
73. D. A. Brodie, P. G. Cook, B. J. Bauer, G. E. Dagle, *Toxicol. Appl. Pharmacol.* 17: 615 (1970).
74. R. Güller, *Schweiz. Med. Wochenschr.* 117: 1527 (1987).
75. H. J. Kaufmann, H. L. Taubin, *Ann. Intern. Med.* 107: 513 (1987).
76. A. G. Schwartz, S. T. Targan, A. Saxon, W. M. Weinstein, *N. Engl. J. Med.* 306: 409 (1982).

77. D. J. Pearson, N. A. Stones, S. J. Bentley, *Br. Med. J.* 287: 1675 (1983).
78. P. A. McCormick, D. O'Donoghue, B. Lemass, *Ir. Med. J.* 78: 17 (1985).
79. S. P. Marcuard, M. N. Ehrinpreis, W. F. Fitter, *J. Rheumatol.* 14: 142 (1987).
80. B. Kirkham, L. Wedderburn, D. G. Macfarlane, *Br. J. Rheumatol.* 28: 272 (1989).
81. H. E. Langer, G. Hartmann, G. Heinemann, K. Richter, *Ann. Rheum. Dis.* 46: 787 (1987).
82. C. W. Jackson, N. Y. Haboubi, P. J. Whorwell, P. F. Schofield, *Gut* 27: 452 (1986).
83. A. G. Fam, T. W. Paton, C. J. Shamness, A. J. Lewis, *J. Rheumatol.* 7: 479 (1980).
84. R. Eaves, J. Hansky, P. Wallis, *Aust. N.Z. J. Med.* 12: 617 (1982).
85. W. C. Fortson, F. J. Tedesco, *Am. J. Gastroenterol.* 79: 878 (1984).
86. D. M. Martin, J. A. Goldman, J. Gilliam, S. M. Nasrallah, *Gastroenterology* 80: 1567 (1981).
87. P. Gillet, B. Bannwarth, J. F. Chambre, P. Pere, A. Gaucher, *Thérapie* 44: 301 (1989).
88. M. Chaffman, R. N. Brogden, R. C. Heel, T. M. Speight, G. S. Avery, *Drugs* 27: 378 (1984).
89. R. Behrens, M. Deveraux, B. Hazleman, K. Szaz, J. Salvin, G. Neale, *Gut* 27: 59 (1986).
90. J. Hardcastle, P. T. Hardcastle, D. K. Kellher, *J. Physiol.* 354: 85P (1984).
91. H. V. Ammon, S. A. Fowle, R. A. Komorowski, *Clin. Res.* 32: 537A (1984).
92. U. Bross-Bach, J. G. Saal, F. Hartmann, C. A. Müller, H. D. Waller, *Z. Rheumatol.* 46: 201 (1987).
93. C. J. Michet, J. Rakela, H. S. Luthra, *Mayo. Clin. Proc.* 62: 142 (1987).
94. D. Jarner, A. Mertz Nielsen, *Scand. J. Rheumatol.* 12: 254 (1983).
95. E. Beubler, in *New Trends in Pathophysiology and Therapy of the Large Bowel* (L. Barbara, M. Miglioli, S. F. Phillips, eds.) Elsevier, Amsterdam (1983) p 119.
96. F. J. Tedesco, J. T. DiPiro, *Am. J. Gastroenterol.* 80: 303 (1985).
97. M. Balazs, *Dis. Colon Rectum.* 29: 839 (1986).
98. H. W. Steer, D. G. Colin-Jones, *J. Pathol.* 115: 199 (1975).
99. B. Smith, *Dis. Colon Rectum* 16: 455 (1973).
100. C. J. C. Roberts, in *Gastrointestinal Disease* (C. J. C. Roberts, ed.) Springer, Berlin (1983) p 169.
101. W. D. Williamson, W. A. Hooge, *J. Can. Assoc. Radiol.* 32:133 (1981).
102. B. Wörmann, W. Höchter, H.-J. Seib, R. Ottenjahn, *Endoscopy* 17: 165 (1985).
103. V. Wienert, E. I. Grussendorf, *Hautarzt* 31: 668 (1980).

104. K. Schaarschmidt, H. J. Richter, E. Gross, F. W. Eigler, *Chirurg* 55: 584 (1984).
105. V. F. Eckardt, G. Kanzler, W. Remmele, *Gastroenterology* 91: 1123 (1986).
106. H. Fromm, M. Malavolti, *Clin. Gastroenterol.* 15: 567 (1986).
107. C. J. C. Roberts, T. K. Daneshmend, in *Gastrointestinal Disease* (C. J. C. Roberts, ed.), Springer, Berlin (1983) p 202.
108. C. P. Denaro, *Rat. Drug. Ther.* Suppl. 22: 1 (1988).
109. F. Halter, *Schweiz. Med. Wochenschr.* 111: 773 (1981).
110. J. Walls, D. Bell, W. Schorr, *Br. Med. J.* 1: 52 (1952).
111. D. L. Woolf, *Br. Med. J.* 1: 1497 (1965).
112. V. Wright, R. A. Hopkins, *Rheumatol. Rehab.* 18: 186 (1979).
113. R. Cheli, G. Ciancamerla, *Minerva Dietol. Gastroenterol.* 20: 56 (1974).
114. R. Ottenjahn, J. Altaras, K. Elster, P. Hermanek, in *Atlas der Darmerkrankungen, Dickdarm I,* (Pharmazeutische Verlagsgesellschaft, München (1983).
115. I. Anthony, M. D. Bravo, M. Robert, M. D. Lowman, *Radiology* 90: 113 (1986).
116. B. Delcambre, A. Bleuse, B. Bodelle, L. Catanzariti, J. J. Delannoy, L. L. Derreumaux, J. C. Farasse, A. Hennebo, X. Lelieur, *Lille Med.* 25: 49 (1980).
117. J. L. Badetti, A. Chambard, C. Gueyffier, *Gastroenterol. Clin. Biol.* 13: 313 (1989).
118. M. B. Robertson, K. J. Breen, P. V. Desmond, M. L. Mashford, A. M. McHugh, *Med. J. Aust.* 1: 243 (1977).
119. R. H. Lusk, F. R. Fekety, J. Silva, T. Bodendorfer, B. J. Devine, H. Kawanishi, L. Korff, D. Nakauchi, S. Rogers, S. B. Siskin, *J. Infect. Dis.* Suppl. 135: S111 (1977).
120. R. E. Condon, M. J. Anderson, *Arch. Surg.* 113: 794 (1978).
121. A. P. Roberts, A. W. Hughes, *Int. J. Orthop.* 8: 299 (1985).
122. J. F. Tures, W. F. Townsend, H. D. Rose, *JAMA* 236: 948 (1976).
123. D. Ruff, J. Jaffe, R. London, J. Candio, *J. Urol.* 134: 1218 (1985).
124. F. J. Tedesco, *J. Infect. Dis.* Suppl. 135: S95 (1977).
125. G. Thomson, A. H. Clark, K. Hare, W. G. S. Spilg, *Br. Med. J.* 282: 864 (1981).
126. M. Dan, Z. Samra, *Am. J. Med.* 87: 479 (1989).
127. X. Roblin, C. Baudry, F. Becot, J. Abinader, D. Monnet, *Presse Med.* 19: 922 (1990).
128. B. J. O'Keefe, G. S. Tillotson, *Lancet* 336: 1509 (1990).
129. N. Low and A. Harries, *Lancet* 336: 1510 (1990).
130. F. Klotz, M. Barthet, M. Perreard, *Ann. Med. Interne* 141: 276 (1990).

Index